The Hunt for Planet X

Kulbaco

Govert Schilling

The Hunt for Planet X

New Worlds and the Fate of Pluto

Copernicus Books

An Imprint of Springer Science+Business Media

ISBN 978-1-4419-2667-8   e-ISBN 978-0-387-77805-1

© 2010 Springer Science + Business Media, LLC

Published in the United States by Copernicus Books,
an imprint of Springer Science + Business Media.

Copernicus Books
Springer Science + Business Media
233 Spring Street
New York, NY 10013
www.springer.com

Translation from Dutch: Andy Brown
Author photo: Hans Hordijk
Springer acknowledges the translation subsidy by
    the NLPVF: Foundation for the Production and
    Translation of Dutch Literature

Dutch edition: De jacht op Planeet X, Fontaine
    Publishers, the Netherlands 2007, ISBN 978-9-059-56194-6

Manufactured in the United States of America.
Printed on acid-free paper.

For Margreet, with whom I shared a fantastic journey of discovery of a completely different kind during the writing of this book

Foreword

In August 2006, Pluto was disqualified as a planet. According to the International Astronomical Union the solar system only has eight planets instead of nine. Pluto, discovered in 1930 during a comprehensive search for the mysterious Planet X, now has the second-rate status of a dwarf planet.

The popular belief that astronomy is a rational science that leaves no room for emotion is a misconception. The news of Pluto's demotion made the front pages of newspapers around the world. Everyone had something to say about it, and emotions had rarely run so high in astronomical circles. But, of course, the new planet classification was founded on scientific fact and insight. In the early 1990s, what had been suspected for many years now proved indisputable – Pluto was only one object in an enormous population of small ice dwarfs in the outer regions of the solar system. One of them even turned out to be slightly bigger.

This book traces the history of exploration and discovery in the outer solar system. Running through that history like a red thread is the hunt for Planet X – a hypothetical celestial body that has remained elusive. That the hunt continues is not surprising: astronomers perpetually speculate on what is just beyond the reach of their telescopes. Every new discovery is incorporated in our evolving picture of the solar system, and each one opens up possibilities for a new Planet X.

Science is practiced by living, breathing people, and that is also true of the search for objects billions of kilometers from the Earth. It is a journey of discovery that has been under way for many centuries, a journey that has been marked by sweat and tears, and sometimes even a little blood. And, just as in any thrilling story of adventure, the hunt for Planet X has seen passion and dedication, success and failure, vanity and suspicion, and – above all – limitless curiosity. *The Hunt for Planet X* is not a dull schoolbook, but a scientific detective story.

And it is a detective story that is played out closer to home than you might think at first glance. The discoveries in the outer solar system may have cost us a planet, but they have provided many new insights into the origin and evolution of the planets. And into our own place in the cosmos.

No area of scientific endeavor can be covered exhaustively in a popular-scientific book. During the writing of *The Hunt for Planet X*, it was necessary to omit certain parts of the story. It was also impossible to devote attention to all those scientists working in the field of solar system research. If that has resulted in an unbalanced presentation of the subject matter, it was not my intention and I take full responsibility.

I owe many thanks to all the astronomers who consented to be interviewed during the research for this book: Mike A'Hearn, Charles Alcock, John Anderson, Fran Bagenal, Rick Binzel, Alan Boss, Mike Brown, Marc Buie, Robin Canup, Andy Cheng, Jim Christy, Dale Cruikshank, Imke de Pater, Martin Duncan, Dan Durda, Julio Fernández, Brett Gladman, Dan Green, Will Grundy, Piet Hut, Dave Jewitt, Nick Kaiser, Charles Kowal, Hal Levison, Jack Lissauer, Jane Luu, Renu Malhotra, Brian Marsden, Bill McKinnon, Bob Millis, Alessandro Morbidelli, Richard Muller, Max Mutchler, Keith Noll, Joel Parker, Venetia Phair-Burney, David Rabinowitz, Paolo Santos-Sanz, Dava Sobel, Myles Standish, Andrew Steffl, Denise Stephens, Alan Stern, Dave Tholen, Chad Trujillo, Tom Van Flandern, Larry Wasserman, Paul Weissman, Iwan Williams, Eliot Young and Leslie Young. Their impassioned accounts have made *The Hunt for Planet X* an exciting read.

Govert Schilling, summer of 2008

Contents

Chapter 1
A Larger Solar System

Jelte Eisinga jumped with alarm as the front door of the house, *De Ooijevaar*, slammed loudly. A moment later his father came storming into the kitchen and angrily threw the latest edition of the *Leeuwarder Courant* onto the table. Not even noticing his 6-year-old son, Eise stamped up the stairs to the attic, still mumbling to himself. Jelte started to cry and, when his mother came in from the yard to see what all the commotion was about, he hid himself in the folds of her dress. Pietje was all too familiar with her husband's moods, but recently, now his life's work was nearing completion, he had seemed to be more content. What could have upset him so much?

Stroking Jelte's back, Pietje picked the newspaper up from the table. Almost immediately, she saw a report that both excited and alarmed her. She felt the blood rush to her head and involuntarily glanced across at the door to the living room, which for many years had practically been forbidden territory to mother and son. In England, on March 13, 1781, William Herschel had discovered a new planet, way beyond the orbit of Saturn. In one fell swoop, the solar system had doubled in size. Pietje thought about the ceiling in the living room, the box bed, the attic and, above all, her husband's passion. She wouldn't survive another 7 years living in all this mess.

Little Jelte had just been born when Eise Eisinga told his wife about his dream: to construct a moving scale model of the solar system. It was the spring of 1774, a relatively calm time in their wool-combing business in the Friesian city of Franeker. Pietje remembered as if it were yesterday how Saturn had twinkled high in the evening sky, in the constellation of Leo. 'Look,' Eise had said, 'that is the furthest planet, with the largest and slowest orbit. That one will have the largest wheel, with 538 teeth.' Saturn's enormous orbit would of course have to fit into the living room, so the scale had to be modified to one in a billion. He was 30 years old, a man with a mission and with his sights set far beyond the horizon. How could she ever have resisted him?

G. Schilling, *The Hunt for Planet X*, DOI 10.1007/978-0-387-77805-1_1,
© Springer Science+Business Media, LLC 2009

ARCES ATTIGIT IGNEAS

Eise Eisinga, painted by Willem Bartel van der Kooi in 1827 (Courtesy Royal Eise Eisinga Planetarium)

So *De Ooijevaar* was transformed into a workplace. Shafts were turned on the lathe, wooden wheels sawn into shape, hoops bent into circles, and nails forged. A ceiling was removed here and a wall knocked down there. At one point, the pendulum of the main clockwork mechanism nearly ended up in their box bed, but Pietje put her foot down just in time. An attic full of cogwheels, circular slits in the ceiling, and more dials on the wall than you could find on the church tower in Franeker – that was all bad enough, especially as they also had to keep the combing business going.

Eise said that he would need 7 years to complete the project and she had to admit, he had kept his promise. But it had been 7 years of sawdust, frustration, and sleepless nights. And years in which the planets received a great deal more attention than little Jelte, who grew up amidst a solar system in the making. And now, at the very time that the final touches had been made to the impressive orrery, the walls painted and the planets polished, with the first curious visitors coming to take a look, this Herschel had made a discovery that had rendered the whole masterpiece obsolete. Pietje could hear Eise stamping around above her head. How awful he must feel! She laid the *Leeuwarder Courant* back on the table and started shelling the peas.

Pea

If you imagine the Earth the size of a large pea, the Moon would be a pinhead 20 centimeters away. The Sun, the size of a beach ball, would be 75 meters away. Somewhere between the Earth and the Sun, there will be two other peas in orbit: Mercury and Venus. And beyond the orbit of the Earth is Mars, the size of a bead. The giant planets Jupiter and Saturn are larger – rubber balls about 6 or 7 centimeters in diameter – and are much further away. Jupiter would be in orbit just under 400 meters from the Sun, while Saturn would be almost 1,500 meters distant. A beach ball, two rubber balls, three peas, a bead, and a pinhead distributed over an area the size of 150 football fields – that is the solar system as Eisinga knew it.

This was much different from the cosmos of the ancient Greeks. They had no idea that they lived on a pea in orbit. They thought the Earth was the hub of the universe, a stationary central point around which the celestial bodies rotated. They knew of seven moving bodies: in addition to the Sun and the Moon, there were the 'wandering' stars (Mercury, Venus, Mars, Jupiter, and Saturn) which traced loop-shaped courses among the 'fixed' stars. Our word 'planet' comes from the Greek word for 'wandering star.'

When, 200 years before Eise Eisinga was born, Polish astronomer Nicolaus Copernicus announced his revolutionary theory that the Sun did not revolve around the Earth but vice versa, the Sun and the Moon were no longer seen as planets. Now, the Earth itself was one of the six planets circling the Sun, halfway between the orbits of Venus and Mars, substantially larger than Mercury, but much smaller than Jupiter and Saturn. And, as four moons had been discovered around Jupiter and five around Saturn since the invention of the telescope in the early seventeenth century, there was also a chance that a new planet might one day be discovered.

Astronomers did regularly discover new comets in the firmament. They were impressive celestial bodies with curved tails, moving through the solar system in strange, elongated paths. No one could have imagined at the time that these mysterious tailed stars were icy remnants of the formation of the solar system. And it is logical that they would strike fear into the hearts of ignorant farmers and everyone else, countryfolk, and townspeople alike. Unexpected phenomena in the heavens, such as eclipses, conjunctions, meteor showers, and comets appeared to mock the order and regularity of the cosmos. They could only mean trouble.

No one was more conscious of that fear than Eelko Alta. And if the people of Friesland were not already afraid of strange cosmic phenomena, it was always possible to make them afraid. Eelko Alta was a clergyman, and there is no more fertile ground for faith and devotion than fear of the Almighty's wrath. It may not, however, have been entirely a matter of opportunism on Alta's part. Shortly after the middle of the eighteenth century, apocalyptic omens were the order of the day and it is not inconceivable that the Friesian clergyman really was convinced that the Day of Judgment was at hand.

In the fall of 1754, at the age of 31, after preaching for 9 years in Beers, Alta was transferred to the small village of Bozum, a few kilometers to the northeast of Sneek. Just over a year later, the world suddenly moved violently. On November 1, 1755, a powerful earthquake set all the water in canals, rivers, lakes, and ponds in Holland and Friesland in motion. Ships were thrown against quaysides and anchor chains broke. One of Alta's older colleagues, Johan Georg Muller, who was a clergyman in Leeuwarden, promptly published a booklet entitled *Voortekenen van de nabijheid van het vergaan der wereld* (Signs that the end of the world is nigh.)

The discovery shortly afterwards of a new comet, and the fact that the Earth moved again on February 18, 1756, must have made a great impression on the readers of Muller's booklet. Was the End of Days really imminent? Was the blood-red eclipse of the Moon on January 24, 1758 a new omen? And what about the comets that appeared in the sky in 1759 and 1760? Or the strange dot that moved across the face of the Sun on June 6, 1761 and which, according to the professors at the University of Franeker, was the planet Venus – but now pitch black instead of white? And what of the ring-shaped solar eclipse on April 1, 1764? Was it perhaps a coincidence that, at the end of the 1760s, there was another mysterious outbreak of disease among Friesian cattle?

Conjunction

On February 19, 1774 Eelko Alta, now 50 years old, opened the *Leeuwarder Courant*. His attention was immediately drawn to a press report from Dresden. German astronomers had calculated that, in the early morning of Sunday, May 8 there would be a rare conjunction of Jupiter, Mars, Venus, Mercury, and the Moon in the constellation of Aries. If that wasn't a portent of the Last Judgment, then what was? Alta's thoughts went back to the earthquake of 1755 and Muller's booklet. That same evening, he started writing and, less than 2 months later, the same *Leeuwarder Courant* announced the publication of his book.

Drawing of the planetary conjunction of May 8, 1774 by P.Y. Portier (Courtesy Royal Eise Eisinga Planetarium)

It was a very long announcement. Not because there was so much to say about the book, but because it had an unusually long title, which certainly deserves to be quoted in full: *Theological and Philosophical Contemplations on the Conjunction of the Planets Jupiter, Mars, Venus, Mercury and the Moon, due to occur on the 8th of May 1774, and about the Possible and Probable Astronomical and Physical Consequences of this Conjunction. From which it can be ascertained that it may have an impact not only on the Globe, but on the entire Solar System, to which we belong, and which could be a preparation for or a beginning of the Dismantling or Destruction thereof, in part or in whole. By a lover of the Truth.*

Eighty-eight pages of doom and disaster, embellished with nearly a hundred Biblical quotes, tried to persuade the reader that May 9 would probably never see the light of day. The Friesland Provincial Executive tried to prevent undue panic by confiscating all the printed copies and not releasing them until May 9, but this did little to put the public's mind at rest. Perhaps Pietje Eisinga–Jacobs was worried too: she was about to give birth, and a year earlier her little daughter Trijntje had died when only a few weeks old. This time she wanted to give Eise a healthy child, so she could do without bad omens from the heavens. But May 8 came and went, the conjunction was hardly visible in the dawn sky and, on May 15, Alta was back in his pulpit in Bozum as usual.

For wool-comber Eise Eisinga the commotion around the conjunction of the planets was the direct cause of his decision to build an orrery. Ever since his boyhood, Eise had shown a keen interest in mathematics and astronomy. At the age of 17, he had observed the transit of Venus with great fascination and had written his first astronomical treatises. Two years later he made detailed calculations of solar and lunar eclipses. Wouldn't it be fantastic if he could build a scale model to show everyone just how innocent an apparent conjunction of the planets really is, and how wonderful the regularity and predictability of the solar system? Three weeks after the conjunction, Pietje gave birth to a healthy son – who they named Jelte, after Eise's father – and he ventured to discuss the idea with his wife. Seven years later, in 1781, his dream came true and today, the Eise Eisinga Planetarium is the oldest working mechanical orrery in the world.

Interior of the Eise Eisinga Planetarium in Franeker, the Netherlands, completed in 1781. On the ceiling is a moving scale model of the solar system (Courtesy Royal Eise Eisinga Planetarium)

Fuzzy Star

The year 1781 was then also the year in which the solar system suddenly doubled in size with the discovery of a new planet. While Eise Eisinga was hard at work in Franeker doing calculations, forging nails, and sawing grooves to capture the firmament in a moving model, across the North Sea in Bath, England, William Herschel could be found every clear night directing his self-built telescopes at that same firmament. He charted star clusters, recorded binary stars, and catalogued nebulae. His telescopes were among the best in the world, so an enormous undiscovered territory lay at his feet – or rather, above his head. It was only a matter of time before Herschel would come across the planet Uranus.

William Herschel, the discoverer of the planet Uranus (From: *Famous Men of Science*, Sarah K. Bolton)

On March 13, 1781, in the constellation of Gemini, Herschel discovered a fuzzy star that changed its position slightly every night. At first, he thought he had found a new comet, but it soon became clear that the object was following a slow, practically circular path around the Sun, far beyond the orbit of Saturn. A new planet had never been found before and, after announcing his find, Herschel immediately became world famous. He received several awards, was elected to the Royal Society and was appointed Astronomer Royal to King George III. This was perhaps partly in recognition of Herschel's generous gesture in naming the new planet *Georgium Sidus* ('George's Star'). The name did not last long, however. French astronomers refused to accept a planet named after an English king and resolutely referred to the new discovery as 'Herschel.' German astronomer Johann Bode solved the problem by proposing the name Uranus, the mythological father of the Roman god Saturn.

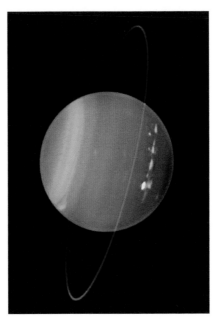

Uranus with its clouds and rings, as seen by the 10-meter Keck Telescope on Hawaii (W.M. Keck Observatory)

Herschel discovered Uranus by coincidence; he had not been searching for a new planet. And, as is often the case with coincidental discoveries, it could have happened much earlier if astronomers had been a little more attentive. In 1690, almost a hundred years before it was discovered, Uranus had been recorded by the English astronomer John Flamsteed. But he thought it was just a regular star in the constellation of Taurus and catalogued it as 34 Tauri. If Flamsteed had noted 34 Tauri's slow change of position, the seventh planet would have been discovered more than 50 years before Eise Eisinga was born. And, who knows, Eisinga's dream of building an orrery might never have become a reality – after all the living room in *De Ooijevaar* would simply never have been big enough.

Pietje and little Jelte did not have to live in a construction site for another 7 years. The orrery was completed according to the original design and it still provides us with an excellent model of the 'classical' solar system. Three years after it was completed, father and mother Eisinga had a second son, Jacobus, and 4 years later, when Jelte was just 14, Pietje died. Jelte himself only lived to be

34, to the great sorrow of his father, who had taught him the basic principles of mathematics and astronomy.

And what about Eise himself? At the end of the eighteenth century, he was forced to live abroad for some time for political reasons. During these years, he dreamed of building a new orrery. Bigger and better, under an enormous planetarium dome. And if he had enough room, he could include the new planet Uranus. Eise's new dream never became reality. The original orrery was bought by the State in 1825, 2 years before Eisinga's death, and transferred to the municipality of Franeker in 1859. It is still to be found in *De Ooijevaar* on what is now known as the Eise Eisingastraat. It is one of the most fascinating astronomical reminders of the turbulent second half of the eighteenth century, when apocalyptic superstition and scientific curiosity vied for prominence and a new planet was found beyond the limits of the known solar system for the first time. It would not be the last.

Chapter 2
Eleven Planets

Giuseppe Piazzi never expected to see the beginning of the nineteenth century. When he was born, on July 16, 1746, his mother had already given birth to eight children, but most of them had died shortly after coming into the world. To assure the salvation of his new-born soul, Giuseppe was christened almost immediately after his birth in the family house in Ponte Valtellina. The quick christening proved unnecessary, but Piazzi was never a healthy man, despite living a devout life as a member of the Theatine Order.

Giuseppe Piazzi, who discovered the first asteroid (Courtesy Stichting De Koepel)

And now here he was, 54 years old, on the evening of the first day of the new century, peering through the eyepiece of the meridian circle in the observatory in Palermo. Thursday, January 1, 1801 was a crystal clear evening on the north

G. Schilling, *The Hunt for Planet X*, DOI 10.1007/978-0-387-77805-1_2,
© Springer Science+Business Media, LLC 2009

coast of Sicily. Venus had just set in the southwest; the red planet Mars twinkled high above the southern horizon. And 9 degrees to the south of Mars, in the 'shoulder of the Bull' – as he himself described it – Piazzi discovered a new planet. A planet which astronomers had been speculating about for nearly 200 years.

Even stranger, it was a planet that had simply been invented, by German astronomer Johannes Kepler in 1596. Kepler not only was as unhealthy as Piazzi, but also suffered from a stutter and was the son of a convicted witch to boot. He was 25 when he published his *Mysterium cosmographicum*, a rather esoteric book in which he sought the divine idea behind the design of the solar system. Why were there 6 planets instead of 7, or 12? What system lay behind their orbital periods and the distances between them? And above all, what reason could God possibly have had for leaving such a large empty space between the orbits of Mars and Jupiter?

Like the Greek philosopher and mystic Pythagoras, Kepler was convinced that nature must be ordered according to mathematical rules. But the divine code of the solar system remained a mystery. As a last resort, he imagined a new planet in the wide, empty zone between Mars and Jupiter. That made the solar system a little more systematic, but it was still not enough.

Partly thanks to Kepler, who in the early seventeenth century established the laws governing the movements of the planets, their orbits were much better documented 200 years later. But the puzzle of the empty space between Mars and Jupiter remained unsolved. And more and more astronomers started to seriously consider the idea that, somewhere in that space, there had to be an as yet undiscovered planet.

That was largely due to an influential book by another 25-year-old German astronomer, Johann Elert Bode. Bode would later suggest the name Uranus for the planet discovered by William Herschel in 1781. He also became the director of the Berlin observatory and, in 1801, published his *Uranographia*, one of the most significant star atlases in history. But his reputation was already secured when he wrote *Anleitung zur Kenntnis des gestirnten Himmels* in 1772.

In that book Bode described a remarkably simple formula. Take the numerical series 0, 3, 6, 12, 24, 48, and 96, where each number is double that of the one before it. Add four to each number, so that you get 4, 7, 10, 16, 28, 52, and 100. Divide the result by 10 and you get approximately the distances of the planets from the Sun in astronomical units, where one astronomical unit (AU) is equal to the distance from the Earth to the Sun (150 million kilometers). It was amazingly simple, but there was one problem: the formula was only accurate if the fifth number (2.8) was reserved for a planet between the orbits of Mars (at approx. 1.6 AU) and Jupiter (approx. 5.2 AU).

Quest

Bode did not discover this mathematical rule himself. Six years previously, his compatriot Johann Daniel Titius had described it in a very free translation of a French natural-scientific book. Titius, in his turn, had taken it from the German Christian Wolff, who had mentioned it in 1724, and Wolff was probably familiar with the work of the Englishman David Gregory, who had drawn attention to the extraordinary ratios between the orbits of the planets in 1702. In effect, the idea of an unknown planet between Mars and Jupiter had never really been out of fashion since Kepler's time.

But 'Bode's Law' was not really taken seriously until the discovery of Uranus. Uranus orbits the Sun at an average distance of 2.87 billion kilometers – just over 19 AU. And the eighth number in the series identified by Titius and Bode is 19.6 (192 plus 4 divided by 10)! That could of course not be a coincidence. At the end of the eighteenth century, Bode and his contemporaries were practically convinced that there must, indeed, also be another planet between the orbits of Mars and Jupiter. The big question was: how to find it?

It had to be a small, dark world otherwise it would have been observed long ago with the naked eye. There was nothing else for it but to search the heavens with a telescope for an inconspicuous speck of light moving between the stars. There was one plus point: all of the known planets move around the Sun in more or less the same plane and therefore, seen from the Earth, are always found in one of the 12 constellations of the zodiac. The quest could therefore be restricted to a relatively narrow zone.

The Hungarian nobleman Baron Franz Xaver von Zach, chief astronomer to the Duke of Gotha and director of the observatory in Seeberg, was determined to find the new planet, but soon realized that he could never do it alone. It would need a joint effort – and preferably an international one.

On September 20 and 21, 1800, six astronomers came together at the observatory in Lilienthal to set out a strategy. In addition to Baron von Zach, they were Johann Schröter, Karl Harding, Heinrich Olbers, Ferdinand von Ende, and Johann Gildemeister. They divided the zodiac into 24 equal parts and drew up a list of an additional 18 European astronomers who would join them in the detective work.

Later that year, they sent out the first letters to their selected colleagues in Germany, Denmark, Sweden, Russia, France, Austria, England, and Italy, asking them to take part. It was the most ambitious program of cooperation ever in the history of astronomy. There could be no doubt that the *Himmelpolizei* (Celestial Police), as Von Zach called the group of 24 investigators, would track down the unknown planet.

Point of Light

Giuseppe Piazzi knew nothing of all this when, early in January 1801, he discovered a small, moving point of light about 9 degrees to the south of Mars. He had been on the list drawn up by the small group in Lilienthal, but for some reason or another he had never received Von Zach's letter. He had heard of the theories about a planet between Mars and Jupiter, but only in passing. After all, Piazzi was a mathematician and not an astronomer. He had not taken much interest in astronomy until 1787, when the Royal Academy in Palermo asked him to set up an observatory.

Piazzi had commissioned one of the best telescope-builders in the world, Jesse Ramsden in London, to build a splendid meridian circle for the new observatory. Meridian circles, which are always directed exactly to the south, are exceptionally suited to charting the positions of the stars. Piazzi used the new instrument to produce an accurate stellar catalogue, which would eventually be published in 1803. It was with this telescope that, on January 1, 1801, he measured the position of a large number of faint stars near the zodiacal constellation of Taurus.

It was not until the following day, when Piazzi made a second series of measurements that he realized that one of the stars had moved slightly in his field of vision. He thought that perhaps he had made an error on the first evening, but on January 3 the star had moved slightly again, and again the following night. It all seemed to point to the discovery of a new comet, but there was no sign of the nebulosity that generally surrounds comets.

Once Piazzi had assured himself that he was not mistaken, he announced the discovery of the new comet. He did not, however, give its position: for the moment, at least, it was 'his' find. Once he had made more observations, he could probably calculate the object's path through the solar system before someone else beat him to it. He was still not sure that the sharp little point of light really was a comet. Perhaps he had made an even more important discovery. At the end of January, Piazzi wrote to his good friend and colleague Barnaba Oriani in Milan: 'I have announced this star as a comet, but since there is no sign of the usual nebulosity and it moves in a very slow and even manner, I suspect that it may be something more significant than a comet.'

Piazzi did not have much luck with the follow-up observations. In January, the weather was often bad and in February, the new star was already in the south at the end of the afternoon, as the Sun set. It was impossible to conduct positional measurements with the meridian circle during twilight. To make matters worse, Piazzi became ill. The long nights staring through the telescope had affected his already weak health. By the end of the winter, he had gathered only 24 measurements.

Meanwhile, other astronomers were starting to get impatient. An important new discovery was all very well, but they wanted to see evidence. Piazzi, who was still trying to calculate the object's path, did not want to release his findings yet but, under pressure from the French astronomer Joseph Lalande, he could

Ceres, as imaged by the Hubble Space Telescope (NASA, ESA, J. Parker (Southwest Research Institute), P. Thomas (Cornell University), L. McFadden (University of Maryland, College Park), and M. Mutchler and Z. Levay (STScI))

hardly refuse. As well as being a Theatine monk, brother Giuseppe was also a freemason, and Lalande was Grand Master of the French Lodge of the Nine Sisters, and therefore not someone to argue with. In April, Piazzi sent his findings to Paris, with a copy to Oriani in Milan, but not to Johann Bode in Berlin.

Bode, who was now the director of the observatory in Berlin, had received an earlier letter from Piazzi about the 'new comet' and was actually quite certain that it was the planet that Baron von Zach's *Himmelpolizei* were looking for. Twenty years after he had proposed the name Uranus for the new planet beyond the orbit of Saturn, Bode now came up with the name Juno for the new planet between Mars and Jupiter. And in April, he announced the discovery of the new planet to the press.

Piazzi was furious. How dared the Germans announce the discovery of *his* planet and give it a name? Who had discovered it in the first place? Piazzi himself had thought of calling it Ceres Ferdinandea, after the goddess of Sicily and King Ferdinand IV, the ruler of the island. Fortunately he had the support of Von Zach, who found Ceres an appropriate name, though he thought the addition of 'Ferdinandea' was a little excessive.

In the meantime, astronomers in Paris had been working hard analyzing Piazzi's positional measurements and there appeared to be no more doubt at all: the wandering star that Piazzi had observed in January and February was not a comet but a planet in a more or less circular orbit around the Sun. But Lalande and his colleagues were unable to determine Ceres' orbit with any great degree of accuracy. Consequently, no one knew where the small planet would appear in the summer months, when it would once again be visible in the morning sky. At the end of 1801, Ceres had still not been rediscovered and Lalande even began to doubt its existence. After all, he had only a series of numbers. Piazzi was the only one who claimed to have actually seen the elusive object.

The orbit problem was finally solved in elegantly by the mathematical genius Carl Friedrich Gauss. At the age of 24, Gauss had developed a method, which astronomers still use today, of calculating the path of a celestial body from a small number of positional measurements. Using Gauss' predictions, Ceres was found again in December 1801 by Von Zach and, independently from him, by Heinrich Olbers on January 2, 1802 – a year after Piazzi's original observations.

The discovery of the eighth planet in the solar system was a fact. Admittedly, it was a small, faint planet, but just like Uranus – discovered 20 years previously – Ceres complied almost perfectly with Bode's Law. And, as a result, the solar system seemed finally to be losing some of its mystery.

Fragments

Heinrich Olbers, the discoverer of Pallas (Courtesy George Beekman)

There was therefore great alarm and confusion when, on March 28, 1802, Olbers discovered another planet between the orbits of Mars and Jupiter. Pallas, as the new planet was named, is about the same distance from the Sun as Ceres but its orbit is a little more elongated and is at a slightly more oblique angle with respect to the orbits of the other planets. Olbers reached a logical conclusion: were Ceres and Pallas perhaps fragments of a former planet? Could that explain why they were so small? And were there other fragments waiting to be discovered in this part of the solar system?

It was indeed curious that, even in the most powerful of telescopes, the new celestial objects remained little points of light, like stars, while all other planets were resolved as small, round disks. This could only mean one thing: Ceres and Pallas could not be more than a few hundred kilometers in diameter. And if more fragments were found, it would be quite illogical to call them all 'planets.' On May 6, 1802, in a presentation at the Royal Society in London, William Herschel therefore suggested referring to Ceres and Pallas forthwith as asteroids ('star-like bodies'). And in a letter to Piazzi he congratulated him on being the first to discover a whole new class of celestial object.

Perhaps the Englishman Herschel was not too keen to share his unique status as the discoverer of a new planet with an Italian and a German. Either way, it showed great insight on his part to derive a complete new population of celestial objects on the basis of observations of only two.

Herschel's proposal found little support among his European colleagues, who enthusiastically welcomed Pallas as the ninth planet. And when Karl Harding discovered Juno on September 1, 1804 and Heinrich Olbers Vesta on March 29, 1807, everyone was in their seventh heaven. In a little over a quarter of a century, no fewer than five new planets (including Uranus) had been discovered – and the *Himmelpolizei* would undoubtedly find more.

But that proved easier said than done. A fifth object (Astraea) was not discovered between the orbits of Mars and Jupiter until December 1845. Neither Herschel, Piazzi, Bode, Von Zach nor Olbers lived to see it. For nearly 40 years, the solar system had 11 planets, which were listed as such in all popular and professional astronomy books published in the first half of the nineteenth century. Starting with the planet nearest to the Sun, they were Mercury, Venus, Earth, Mars, Vesta, Juno, Ceres, Pallas, Jupiter, Saturn, and Uranus. The newcomers were even given stylized symbols like those that have been used to designate the 'classical' planets for many centuries.

After Astraea was discovered, things started to move much faster. Hebe, Iris, and Flora were discovered in 1847, Metis in 1848, and Hygiea in 1849. Five years later, the count had reached 30, and by 1868, no less than 100 objects had been found between the orbits of Mars and Jupiter. They were referred to as minor planets or – as Herschel had proposed – asteroids. And although Ceres, Pallas, Juno, and Vesta continued to enjoy a kind of *status aparte*, everyone realized that it had been a mistake to classify them as full-fledged planets.

The fact that astronomers accepted the 'demotion' of Ceres, Pallas, Juno, and Vesta with such little fuss was not only because those who had discovered them were no longer alive. The most important reason was undoubtedly that, in 1846, a *real* new planet was discovered. And this time it was not by accident. On the contrary.

Chapter 3
The Writing Desk Planet

Elaine Mac-Auliffe had no idea that what she held in her hands was the history of the discovery of the planet Neptune. It was October 1998 and, together with her colleague Nicholas Suntzeff, she was searching the house of Olin Jeuck Eggen, a staff astronomer at the Cerro Tololo Inter-American Observatory in Chile, where Mac-Auliffe was an executive secretary. In a hall cupboard near the bedroom, Elaine and Nicholas found a couple of cardboard boxes containing old books, letters, and files which, according to the catalogue stamp on the title pages, were the property of the Royal Greenwich Observatory in England. One of the bound piles of paper was entitled *Papers relating to the Discovery, Observations and Elements of Astrea, Neptune, Hebe, Iris, Flora. Colors of Astrea, Neptune.* When Greenwich archivist Adam Perkins received an email from Mac-Auliffe about the unexpected find, he jumped for joy. The Neptune File had finally been found.

Eggen was a renowned astronomer who had worked at the Royal Observatory from 1956 to 1961. The observatory had then been housed at the stately Herstmonceux Castle in southern England and Eggen lived in one of the castle rooms and had access to the library day and night. That was very convenient, as he had a great interest in the history of astronomy and was working on short biographies of George Airy and James Challis, two astronomers who had played key roles in the farcical drama surrounding the discovery of Neptune in 1846. Eggen would certainly have regularly consulted the Neptune File, a unique collection of documents about the discovery of the new planet put together by Airy. And it would have been easy for him to keep the book – and perhaps many others – in his own room.

Where is the line between borrowing something, forgetting to return it, and intentionally keeping it? Eggen will never be able to answer that question: he died of a heart attack on October 2, 1998. But a number of things are certain. In 1961 Eggen quarreled with Richard Woolley, the director of the observatory, and left embittered for California. When he left, he took with him a number of boxes containing valuable historical books. They later went with him to Australia, where he was the director of Mount Stromlo Observatory for 11 years, and then in 1977 to Chile. And when the Greenwich library asked him if he had taken the missing Neptune documents, he denied having them.

G. Schilling, *The Hunt for Planet X*, DOI 10.1007/978-0-387-77805-1_3,
© Springer Science+Business Media, LLC 2009

Olin Eggen (Olin J. Eggen photo collection at the AURA Observatory, Courtesy Elaine Mac-Auliffe)

After Olin Eggen's sudden death, Elaine Mac-Auliffe was initially searching his office and house on the observatory campus in La Serena for the addresses of family members. But of course, there were also piles of scientific journals and books to be catalogued. That these included historical books from one of the most famous scientific libraries in the world was remarkable enough. But it was even more surprising to find that the hoard also contained unique documents that would throw new light on the row between the English and the French about the discovery of Neptune.

Orbital Deviations

Ever since the discovery of Uranus in 1781 there had been speculation about the existence of other planets even further away from the Sun. This was not surprising: once one new planet has been found that is so far away that it can only be seen with a telescope, it is logical to assume that there will be more. Furthermore, Uranus complied perfectly with Bode's Law, so why shouldn't there be a planet somewhere in the region of the next number in the series?

But there was an even more compelling reason to believe in the existence of a celestial body beyond the orbit of Uranus. The new planet might conceal itself like a thief in the night, but it left clear tracks in the solar system. At least, in the first half of the nineteenth century, that was the accepted explanation for an unsolved mystery: the deviations in Uranus' orbit.

Isaac Newton's laws of gravity enable us to calculate accurately how a planet circles the Sun. In empty space there is no friction and a planet never suffers from a headwind, so gravity is the only force affecting its orbit. Shortly after the discovery of Uranus, therefore, it was possible to calculate where the planet would be in the night sky in the coming months and years.

There was only one problem: Uranus refused to follow the timetable. Only a few years after it had been discovered, it became clear that its actual position increasingly deviated from the predictions. This is despite the fact that the calculations had been made using John Flamsteed's observations from 1690, which would be expected to quantify the planet's orbit more accurately.

Yet, even when astronomers included the minute gravitational effects of Jupiter and Saturn in their calculations, they could not get Uranus to play according to the rules. This complicated calculation method was first developed by the mathematician Pierre Simon Laplace, who was considered by his compatriots as the Newton of France. In 1808, Laplace's assistant, Alexis Bouvard, published new tables with the future positions of the planets, taking account of all the reciprocal effects, but Uranus continued to go its own way.

All kinds of explanations were put forward. Perhaps Uranus had recently had a collision with a comet. Or perhaps there was a sort of ether in the cosmos, which slowed the planets down a little. But in the 1830s, more and more astronomers became convinced that there could only be one feasible explanation for the deviant behavior of Uranus: there must be another planet somewhere in the outer regions of the solar system.

Mathematical Genius

If you know exactly where all the planets are and you know how massive they are, you can use Laplace's method to calculate how they can disturb each other's orbits. But at a time when such calculations had to be made at a writing desk, using only paper and ink and logarithm tables, it was certainly no easy task. And the other way around – using measured orbital deviations to calculate the position of an unknown planet that may be responsible for them – is of course even more complicated. George Biddell Airy actually believed that it was not possible, at least not on the basis of positional measurements that covered only a small part of the orbit of Uranus. Airy had been Astronomer Royal at the famous Greenwich Observatory since the summer of 1835 and was an expert on planetary orbits. In 1832 he had written a book on recent developments in astronomy, in which he said that Uranus was one of the great unsolved mysteries of astronomy.

John Couch Adams (Courtesy George Beekman)

It was this book that attracted the attention of John Couch Adams, a mathematical genius from Cornwall. Adams, who probably suffered from Asperger's syndrome, conducted precise calculations of solar eclipses as a teenager and won every conceivable mathematics prize at the University of Cambridge. In the summer of 1841, shortly after his 21st birthday, he came across Airy's book in Johnson's bookshop in Cambridge and decided that he would solve the Uranus mystery himself. The unknown planet must and would be found.

Not that Adams shut himself away in his study that very evening and could think of nothing else. The new planet was a long-term project for his rare free evenings and holidays in Cornwall. It took him until 1845 to find the solution. Yes, he concluded, Uranus' orbital deviations could be explained by the presence of a new planet, which would be circling the Sun in a somewhat elongated orbit. From the Earth, it would have to be visible somewhere in the constellation of Aquarius.

John Adams was very short-sighted and did not own a telescope. But the University of Cambridge had its own observatory with a splendid 30-centimeter telescope, designed by Airy in the 1830 s. Through this Northumberland Telescope (named after the Duke of Northumberland, who donated the lens to the university) the new planet should be visible as a tiny disk.

When Airy left for Greenwich in 1835 – 4 years before Adams arrived in Cambridge – he was succeeded as director of the observatory by James Challis. But Challis had better things to do than chase after a hypothetical planet that existed only in the mind of a short-sighted theorist. Although he did not doubt Adams' mathematical talent, he saw little point in spending weeks tracking down all the stars in Aquarius.

Adams had no more success with Airy. Challis had been kind enough to write an introductory letter to the now renowned Astronomer Royal, but when the timid mathematician turned up at Greenwich unannounced in October, Airy was not there. Adams left his visiting card with the butler, together with a sheet of paper summarizing the results of his calculations, and returned unhappily to Cambridge. Two weeks later he received a letter from Airy, but it was also disappointing. It did not contain a promise to set the best telescopes in England to work to look for the new planet, but only a request for more information about a specific aspect of his calculations. Adams did not bother to respond.

Not on the Map

Adams' French colleague Urbain Jean-Joseph Le Verrier was a lot more determined. But, there again, Le Verrier was by no means timid. On the contrary, he was generally considered arrogant. He was 8 years older than Adams, had a prestigious job at the École Polytechnique in Paris, and had become interested in the Uranus mystery in the summer of 1845. Three months later, he presented his first provisional calculations to the French Academy of Sciences, and in the spring of 1846, he made a quite detailed prediction of the new planet's position. It should be somewhere in the border area between the constellations of Capricorn and Aquarius, and the small disk should be clearly visible amongst the countless little points of light that were the stars.

Adams had also not been idle. He had continued to modify and – hopefully – improve his calculations, with the result that he continually came up with different predictions. It was no wonder that James Challis still did not feel the need to instigate a hunt for the elusive planet. He only changed his mind when Airy himself urged him to use the Northumberland Telescope to try and find it. After all, you cannot refuse a request from the Astronomer Royal. At the end of July the search for the writing desk planet finally started in earnest.

Airy was in a hurry. He was aware of the impressive work being done by Le Verrier, but had not told his colleagues in France of Adams' predictions. If the new planet really existed it should of course be discovered in England rather than in France. But Challis' search progressed with difficulty, partly because of

Urbain Jean-Joseph Le Verrier (Courtesy George Beekman)

clouds and excessive moonlight, not to mention yet another new prediction from Adams, but also because of his own carelessness, undoubtedly brought about by his lack of motivation. It later emerged that he must have had the new planet in his sights on July 30 and August 12, but it had not rung any bells. By the middle of September, there were still no results from Cambridge.

In France, Le Verrier was not having much better luck. A short search at the observatory in Paris was called off at the beginning of August by director François Arago, who preferred to spend the valuable telescope time on less speculative observing programs. A new appeal by Le Verrier to the Academy of Sciences on August 31 was also unsuccessful. The French astronomers would undoubtedly have responded more enthusiastically if they had known that, in the deepest secrecy, their colleagues in Cambridge were also searching for the new planet.

On September 18, in desperation, Le Verrier wrote a letter to Johann Gottfried Galle, a 34-year-old astronomer at the Berlin observatory who had sent him a copy of his thesis the previous year. He suggested that perhaps the new planet could be found with the observatory's famous Fraunhofer Telescope. He should of course have directed his request to Johann Franz Encke, who had succeeded Bode in 1826 as the director of the observatory. But Le Verrier had had bad experiences with observatory directors and it seemed more advisable to him to try and rouse the interest of an experienced observer.

And so it happened that, on the evening of Wednesday, September 23, 1846, Galle opened the dome of the observatory and pointed the 22-centimeter telescope at the area between Capricorn and Aquarius (where that year Saturn also twinkled brightly). Earlier that day, he had received Le Verrier's letter. Encke had given his permission to use the telescope for the search that evening. The director himself had better things to do – it was his 55th birthday. So, together with 24-year-old astronomy student Heinrich d'Arrest, Galle set to work.

It was a laborious job. Star for star, they compared the night sky with a brand-new star atlas – Johann looking through the telescope and Heinrich bent over the map. But it did not take them very long. Around midnight, after Galle had passed on yet another position to his student, d'Arrest shouted back enthusiastically: '*Not* on the map!' The following evening, the unknown 'star' – a little over a degree to the northeast of Saturn – had moved a short distance, just as you would expect from a planet beyond the orbit of Uranus.

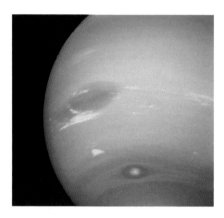

The planet Neptune, as imaged by NASA's Voyager 2 spacecraft in August 1989 (NASA/JPL)

Loss of Face

The news of the discovery swept through Europe like wildfire. Encke called it a 'brilliant discovery.' Heinrich Schumacher, the editor of the authoritative *Astronomische Nachrichten*, spoke of a 'triumph of theory.' Le Verrier was in his seventh heaven. 'Thanks to you, we now have a new world,' he wrote to Galle. The French astronomer and mathematician became a national hero overnight and was overwhelmed with letters of congratulation from leading astronomers across the continent.

But not from England. George Airy was away traveling. John Adams was not at all pleased that Le Verrier and Galle had beaten him to it, and James Challis discovered to his dismay that he had seen the new planet twice without registering what it was. This was a great loss of face for English astronomy. And that is what John Herschel – the son of the discoverer of Uranus William Herschel – must have thought, too. In the weekly *The Athenaeum*, he was the first to write in public about Adams' work and proposed that the young mathematician from Cambridge should also share in the honor of the discovery.

Herschel's revelation received a cool welcome on the other side of the Channel. No one knew who Adams was and Herschel seemed to imply that Le Verrier's calculations were less reliable without independent confirmation. And what was perhaps even worse: both Berlin and Cambridge suggested names for the new planet. Galle thought Janus was an appropriate name, while Challis proposed Oceanus. But who had actually made the discovery possible? Urbain Le Verrier had suggested the name Neptune to the French Bureau des Longitudes, but he was starting to wonder if it could better just be called Le Verrier. In the end, Neptune proved the most acceptable name for everyone.

When Airy returned to Greenwich in the middle of October an international row was in full flow. French newspapers accused the English of 'planet stealing.' The fact that Adams' calculations and Challis' detective work had always been kept a secret did not help. And – Airy had to be honest – if he had been a little more alert a year previously in responding to the strange mathematician who turned up on his doorstep with a piece of paper full of calculations, the planet may have been discovered much earlier – and by an Englishman.

England's honor could only be saved by accepting a certain loss of face. At a crowded meeting of the Royal Astronomical Society in London, on Friday, November 13, Airy, Challis, and Adams gave lectures on what had happened over the past year. Airy had to admit that he had made mistakes, Challis was generally seen as an incompetent observer, while Adams was of course the hero of the day. And although the conflict between France and England continued for several weeks – especially in the press – Airy eventually managed to create the impression that Adams had played at least as important a role in the discovery of Neptune as Le Verrier.

Adams and Le Verrier actually got on very well when they met each other for the first time in England in June 1847. Le Verrier succeeded Arago as director of the Paris Observatory in 1854, while Adams took over from Challis at Cambridge in 1861. Although Le Verrier ran his observatory like a tyrant and was forced to resign by his own staff in 1870, there was never a cross-word spoken between him and Adams.

Le Verrier got his old job back in 1872 when his successor Charles Delaunay drowned during a boat trip. Five years later, on September 23 (the anniversary of the day on which Neptune was discovered), he died at the age of 66. Airy and Adams died within a short time of each other in January 1892, and 3 years later a plaque was unveiled in Westminster Abbey bearing the text: JOHANNES COUCH ADAMS, Planetam Neptunum Calculo Monstravit MDCCCXLV' ('John Couch Adams, whose calculations proved the existence of the planet Neptune in 1845').

Search

Airy's efforts seem to have been successful. After the death of the last of the main characters involved in the Neptune soap opera (Challis died in 1882, Galle in 1910) the world appeared to be convinced that John Adams and Urbain Le Verrier deserved equal credit for the discovery of Neptune, a 'triumph of theoretical astronomy.' Some even felt that Adams deserved greater recognition, since he had started making his calculations before his French colleague. Almost no one knew that Adams' calculations were much less accurate than those of Le Verrier that he continually modified his results and that, because of that, he had been unable to make much of an impression on Challis and Airy.

That information was safely locked away in the archives of the Royal Greenwich Observatory. Airy had collected together all the letters, notes, and newspaper clippings on the discovery of Neptune, but this Neptune File continued to be shrouded in secrecy. Twentieth century historians often had their requests to view the documents refused. And from the early 1960 s, the Neptune File even appeared to be officially missing. Had the intellectual successors of George Airy something to hide?

It was not until October 1998, when Elaine Mac-Auliffe cleared out Olin Eggen's hall cupboard in La Serena that Airy's collection of documents resurfaced. Together with all the other 'borrowed' books, the Neptune File was shipped to England. A year later the documents that everyone thought were lost provided the main source for the book *The Neptune File* by the British science writer Tom Standage, who resolutely insists that his compatriot John Adams played the decisive role in the discovery of Neptune.

Science historian Nick Kollerstrom of University College London has no time for this British chauvinism. 'Airy and his contemporaries have always given the impression that the positions predicted by Adams were only one degree away from those predicted by Le Verrier,' he says. 'But even in the most favorable scenario, Adams' predictions were inaccurate by 4 degrees.' He also doubts the authenticity of the sheet of paper with calculations that Adams supposedly left at Airy's house in the fall of 1845. 'Airy did his utmost to present the facts in a much more favorable light than how it was in reality,' says Kollerstrom, who has made practically the entire Neptune File accessible to everyone through the internet.

The truth will probably never come to light. More than 150 years after the discovery of Neptune, the row about the writing desk planet has become a footnote in the history of astronomy and has long been overshadowed by more recent controversies and intrigues. If one thing is clear from the whole Neptune affair, it is that the discovery of a new world is more than just an interesting scientific achievement. Anyone looking for Planet X finds themselves in a quagmire of ambition, emotion, and envy. And it remains that way today. There is nothing new under the Sun.

Chapter 4
'I've Found Your Planet X'

At first, Carl Lampland didn't notice anything out of the ordinary, on that Tuesday afternoon. It was getting on for 4 o'clock; the working day was nearly over. Through the window of his workroom, in the grounds of Lowell Observatory, he could see the sharp lines of the fir trees set against the cold, bright blue sky. It was February 18, 1930 and, for some reason, it was quieter than normal. Something was missing.

Suddenly, Lampland knew what it was. The ticking had stopped. Throughout the whole afternoon, Clyde Tombaugh had been working with the blink comparator in the room across the hallway. The machine gave a loud, dry click every second, like the metronome for a slow death march. But now it was silent – and had been for a while. As vice director of the observatory, perhaps Lampland should have checked whether the youngest member of his staff was neglecting his duties – Tombaugh used the blink comparator to search photographs of the night sky for an unknown planet. But Lampland's thoughts drifted off. The afternoon silence was rather pleasant. He returned to his work.

About 4.45 he was startled by the sound of Tombaugh's voice. 'Doctor Lampland, I think I've found a *trans*-Neptunian planet.' The vice director jumped from his chair and hurried to the dusky room where the blink comparator stood. So that was why the ticking had stopped so suddenly: Tombaugh had found a small star that moved in the sky and had clearly spent the past three-quarters of an hour verifying his discovery. Lampland sat at the double eyepiece of the machine, purchased by the then director Percival Lowell on Lampland's advice 19 years previously, and was soon convinced. He turned to Tombaugh: 'Time to inform Slipher, Clyde.'

Vesto Melvin Slipher was 54, 2 years younger than Lampland. But he had come to Flagstaff a year earlier, in 1901. He had established a good reputation for himself in the world of astronomy and had succeeded Lowell as director at the end of 1916. Clyde Tombaugh was 24 and had only been working at the observatory for a year. Slipher himself – the man who acquired world fame with his measurements of the recession velocities of galaxies and the expansion of the universe – had hired him especially to search for a planet beyond the orbit of Neptune.

G. Schilling, *The Hunt for Planet X*, DOI 10.1007/978-0-387-77805-1_4,
© Springer Science+Business Media, LLC 2009

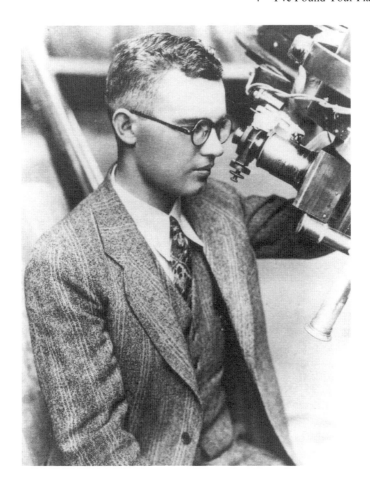

Clyde Tombaugh, the discoverer of Pluto (Courtesy George Beekman)

Tombaugh took a deep breath, pushed his glasses up on his nose, straightened his jacket, walked down the corridor, and knocked on the door of the director's room – an excited farmer's son from Kansas who had discovered a new world. Without waiting for an answer, he opened the heavy door and took a step into the room. Looking the surprised Slipher directly in the eye, he said: 'Doctor Slipher, I've found your Planet X.'

Another Planet?

It was Urbain Le Verrier who had said on September 30, 1846 – a week after the discovery of Neptune – that the 'triumph of theory' might be repeated. Unknown planets tend to betray their presence by their gravitational effects;

anyone who can do basic mathematics can work out where such a planet should be found. Who knows, perhaps Neptune would also prove to have orbital deviations, which would show the way to the next planet.

Neptune, however, moves very slowly: since its discovery nearly a century and a half ago, it has still not completed a full orbit of the Sun. It would therefore take a long time for its orbit to be determined in detail. The planet may have been observed as early as December 1612 and January 1613 as a faint star by no one less than Galileo Galilei, the Italian astronomer who was the first person to see the stars through a telescope, but that was not discovered until 1980. Le Verrier and his contemporaries therefore did not expect any deviations in Neptune's orbit to be uncovered until sometime in the twentieth century.

But Uranus was another story. Since it had been discovered in 1781 it had nearly completed an orbit of the Sun, and a large number of positional measurements were available. Furthermore, now that Neptune's exact position was known its gravitational influence could be calculated very accurately. Even better, shortly after Neptune had been discovered, the Englishman William Lassell had found a moon close to the planet, which had been named Triton. The moon's orbital movement enabled Neptune's mass to be determined – an important factor in the calculations. It was therefore now possible to see whether Uranus behaved as had been predicted.

Around 1880, various astronomers concluded that it did not. Even when Neptune's influence was taken into account, Uranus' movement still displayed small deviations, suggesting that somewhere in the outer regions of the solar system there must be yet another planet. Or perhaps even more than one. It was time for a new search.

At the end of the nineteenth century it became clear that the search would not be conducted in the classical manner, with astronomers peering intensely through their eyepieces: photography had made its entrance in astronomy and it offered unprecedented possibilities. A photographic plate could, if necessary, be exposed for several hours, making visible extremely faint stars that could not be seen through a telescope. During this long exposure time, the telescope had to continue to rotate along with the night sky; otherwise all of the stars would appear as stripes of light.

In 1891 the first celestial body was found with the aid of photography. It was the asteroid Brucia, discovered by the German astronomer Max Wolf. A year later the first comet followed and in 1899 William Pickering, of the Harvard Observatory, found a small moon orbiting Saturn, which was given the name Phoebe. Astrophotography was clearly the best way to hunt for a possible ninth planet (the asteroids Ceres, Pallas, Juno, and Vesta had lost their temporary planet status shortly after the discovery of Neptune).

Lowell's Hunt

Percival Lowell at the eyepiece of the main telescope at Lowell Observatory (Lowell Observatory)

Percival Lowell had resolved to find the unknown planet. Lowell was born on March 13, 1855 to a rich aristocratic family in Boston. His slightly younger brother Abbott became president of Harvard University in 1909 and his much younger sister Amy was a poet who won the Pulitzer Prize in 1926. Percival himself had a great interest in the Far East; when he was about 30 he made several trips to Korea and Japan, and he even played an official diplomatic role during a Korean mission to the United States.

But Percival Lowell's interest in the wonderful culture and religions of the Far East was surpassed by his fascination for the unattainable, mysterious world of the solar system. He decided to solve the riddle of the canals on Mars – dead straight dark lines of the surface of the red planet – and he built an observatory on a mountain top to the west of Flagstaff, Arizona, especially for that purpose. The Lowell Observatory on Mars Hill, which was officially opened on June 1, 1894, was for a long time one of the best observatories in the United States.

Lowell was also a gifted speaker and writer. He gave inspiring lectures on the solar system and published popular books on the possibilities of intelligent beings on Mars, who may have dug the network of canals to irrigate the dry planet with melted ice from the polar caps. In the middle of the twentieth century, the canals on Mars were shown to be an optical illusion and it was proved that the red planet was devoid even of any vegetation, but Lowell believed until his death in 1916 that the Earth's closest neighbor was inhabited.

Around his 50th birthday, in 1905, Lowell started to hunt for a planet beyond the orbit of Neptune. The idea was simple: he hired astronomy students to take photographs of the night sky. Each section was photographed twice, with a few days in between. Lowell would then place the photographic plates (which were like negatives with black stars on a transparent background) one on the top of the other and a little to one side, so that each star appeared double. Because an unknown planet would have moved slightly in the intervening period, the two points of light would be further apart from each other than all the other 'twin' stars.

Lowell had resolved to search the entire night sky in this way, or at least the broad band of the zodiac in which the known planets were located, but it was like looking for a needle in a haystack. By 1907 he had discovered only one asteroid (which was appropriately named Arizona) and he gave up the search.

New Searches for Planet X

Yet Lowell could not forget his fascination for Planet X, as he called the mysterious, unknown planet. And it became even more difficult when he discovered, at the end of 1908, that he was not alone in looking for a *trans-Neptunian* planet. William Pickering was a couple of years younger than Lowell, had established a reputation after discovering Saturn's moon Phoebe and publishing the first photographic moon atlas, and was associated with the renowned Harvard College Observatory in Cambridge, Massachusetts, where his older brother Edward was the director. And although Pickering was a close acquaintance, Lowell had never told him about his ambition to discover the ninth planet.

William Pickering (Courtesy Stichting De Koepel)

Pickering referred to the new planet with the letter 'O,' for the simple reason that it followed the 'N' from Neptune in the alphabet. And, like Le Verrier, he had calculated where Planet O should be in the sky. Although he didn't find anything there, it made Lowell realize that he should make his own search more specific if he were to have any chance of success.

In the spring of 1911, a new hunt for Planet X started on Mars Hill in Flagstaff. According to Lowell's calculations, it should be somewhere in the constellation of Libra. Lampland, who was now vice director of the observatory, insisted on the purchase of a blink comparator, which would make the search for a faint, moving point of light much easier. The principle is simple: you slide into the machine two photographic plates of the same part of the night sky but taken a few days apart. While you look through the double eyepiece, a mirror clatters loudly back and forth showing each image alternately every second. A moving star is then immediately noticeable, because it seems to jump from one side to another.

Pickering was still completely in the dark about Lowell's search. And, it seemed, about his own predictions. In 1911 he suggested that there were three more planets beyond Neptune, which he called P, Q, and R. Later they were joined by the hypothetical planets S, T, and U, and slowly but surely Pickering began to lose his credibility. Early in 1915 Lowell felt bold enough to give a lecture at the American Academy of Arts and Sciences on his hunt for Planet X.

And he quietly hoped of course that the ninth planet would be found just before he was due to speak.

In the meantime, he no longer used the observatory's large 1-meter reflector telescope, but a smaller photographic refractor with a diameter of 23 centimeters, on loan from the Sproul Observatory in Pennsylvania. The refractor had a much larger field of view, speeding up the search. But when the telescope had to be returned in the summer of 1916, Planet X had still not been found. In a period of a little over 2 years, nearly a thousand photographic plates had been exposed, revealing 515 asteroids and 700 variable stars, but no new planet.

Lowell died on November 12, 1916 after suffering a heavy stroke and was buried in a small, Saturn-shaped mausoleum in the grounds of the observatory. He would never know that his Planet X had been photographed twice in the spring of 1915, on March 19 and on April 7. But the tiny point of light was much fainter than anyone had imagined and no one noticed it.

The hunt for the unknown planet was not resumed until 2.5 years after Lowell's death. That was due not so much to a lack of enthusiasm on the part of Vesto Slipher, who succeeded Lowell as director, but more to Lowell's widow, Constance. She tried every legal channel possible to lay claim to a large proportion of the capital of her late husband, who had stipulated in his will that more than a million dollars should be used to continue the search of Planet X.

The telescope with which Pluto was discovered (Lowell Observatory)

It was not until 1929 that Lowell Observatory completed the construction of a new, powerful 33-centimeter refractor telescope with a huge field of view. It was the only telescope in the history of astronomy specially designed to search for an unknown planet. In that same year, a young farmer's son from Kansas was hired to expose the photographic plates and put them in the blink comparator to look for one jumping point of light among millions of faint stars.

Tombaugh's Dedication

Clyde Tombaugh was born in 1906, halfway through Lowell's first serious hunt for Planet X. He showed a great interest in the universe from an early age and built his own telescopes. Enthusiastically, he sent his observations on Mars and Saturn to the Lowell Observatory. Slipher and Lampland, who had themselves grown up on a farm, invited him to come and work at the observatory for a trial period. Tombaugh soon became very popular and, in April 1929, was given the thankless job of searching for the elusive planet.

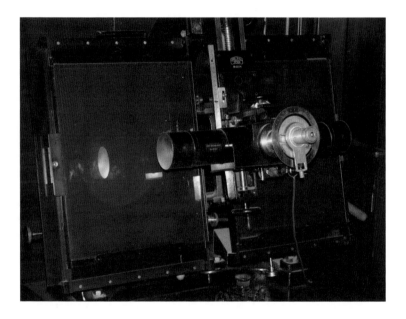

The blink comparator used by Clyde Tombaugh in the search for Planet X (Lowell Observatory)

It was soul-destroying and exhausting work, but Tombaugh was not the type to give up. He developed an efficient search strategy and systematically recorded the entire zodiac. The hundreds of photographic plates that he exposed – each one 33 by 40 centimeters in size – revealed 29,000 new galaxies, 3,969 asteroids, 1,800 variable stars, and two comets. And on the afternoon of Tuesday, February 18, 1930, Tombaugh was looking at two exposures from January 23 and 29, showing a section of the night sky in the vicinity of the star Delta Geminorum, when he saw one of the countless points of light jump a few millimeters back and forth. That was the day when he uttered the words: 'Doctor Slipher, I've found your Planet X.'

Tombaugh had not taken much notice of Lowell's calculations. Lowell had first predicted that Planet X would be found in the constellation of Libra, but changed his mind a few years later, deciding that the search should focus on Gemini. Tombaugh found this all a little vague, and resolved to search the entire zodiac. But when he did eventually find the new planet, it was in Gemini, only 6 degrees from the position predicted by Lowell.

It was therefore no surprise when Percival Lowell received a prominent mention in the telegram announcing the discovery of the new planet several weeks later, once its slow orbital motion had been measured more accurately. The announcement was made on March 13, 1930, Lowell's 65th birthday and the 149th anniversary of the discovery of Uranus. It could hardly have been more fitting. Everything seemed to suggest that the hunt that Lowell had initiated 25 years earlier had finally been successful.

Discovery images of Pluto (arrowed) (Lowell Observatory)

But it did not take long for doubt to set in. Planet X proved to move in an elongated and highly inclined orbit, and was much smaller than had been expected – it was not even visible in the world's largest telescope at the time, the 2.5-meter reflector at Mount Wilson Observatory. It was therefore inconceivable that it was more than six times larger than the Earth, as Lowell had calculated. And if Planet X were much smaller and less massive, how could it be responsible for the remaining unexplained deviations in Uranus' orbit, which had led Lowell to start his search in the first place?

William Pickering – now retired and living in Jamaica – thought he knew the answer. The position of Lowell's Planet X appeared to concur closely with the calculations for his Planet O, which he had published in 1919. And if Planet X were too small to solve the Uranus mystery completely, there might still be hope for his planets P, Q, R, S, T or U. After all, there was no reason at all to assume that there was only one planet orbiting the Sun beyond Neptune. Indeed, shortly after the discovery of Planet X, a Canadian astronomer claimed to have identified another faraway planet on photographs taken in 1924.

Neither this 'Ottawa object' nor Pickering's series of planets have ever been found. But this was no fault of Clyde Tombaugh. After 1930, he continued his search, month after month, year after year. In 1943 – by now a fully trained astronomer, a husband, and a father – Tombaugh had searched the entire night sky, at least as far as it was visible from Flagstaff. That resulted in nearly 800 new asteroids, a handful of galaxy clusters, and one comet. But no new planet beyond the orbit of Neptune.

There seemed to be only one conclusion possible: Planet X, with its strange orbit and small size, really must be the last planet in our solar system. And if this peculiar little object is too small and too lightweight to influence the movement of Uranus, the giant planet's orbital deviations must have another cause and the discovery of Planet X so close to the predicted position must have been sheer coincidence.

Tombaugh died on January 17, 1997, at the age of 90 – an astronomer whose whole life had been determined by one remarkable Tuesday afternoon in February 1930. He had discovered the third new planet in the history of astronomy, and was the first American ever to discover a planet. To the end, he remained convinced that he had completed the inventory of the solar system. Nine planets orbited the Sun, period.

Perhaps it is just as well that Tombaugh never lived to be a hundred. Less than 10 years after his death, the number of planets in the solar system was adjusted again. Not upwards, but downwards, to eight. Henceforth, Planet X – better known as Pluto – had to be content with the status of a dwarf planet.

Chapter 5
The Kid Planet

Venetia Phair-Burney pushes boxes to one side and piles up papers to make space on the small living room table. Outside, on the drive, there is a large 'SOLD' sign. Queensmead, the stately house in Yew Tree Bottom Road, Epsom, just south of London, is full of boxes. Venetia's son Patrick helps her to sort everything out and pack it all up. After the death of her husband in the spring of 2006, it is time to move to a smaller apartment. The Phairs lived in Queensmead for 39 years. The following year they would have celebrated their 60th wedding anniversary. Venetia shuffles through the room – a small but energetic woman with thick, grey curls and friendly eyes behind enormous glasses. Maxwell was 97. She herself is 10 years younger, but she seems determined to reach 100, despite the fact that the death of her husband and the sale of the house have been so exhausting for her.

Venetia Burney was 11 years old when she heard about the discovery of a new planet and thought up the name Pluto, after the Roman god of the underworld. It was Friday, March 14, 1930 and she was eating her breakfast. She grew up in the intellectual environment of the university city of Oxford. Her mother taught at the local primary school and her father was a professor of theology who wrote about the origins and interpretation of the Holy Scriptures. She hardly knew her father: he died in 1925, when Venetia was only six. When her two older brothers went to boarding school, Venetia and her mother Ethel moved in with Grandpa and Grandma Madan in Banbury Road.

Falconer Madan was an exceptional and very distinguished aristocrat. He had been the head of the renowned Bodleian Library at the University of Oxford, and after his retirement he retained an office at the library, to which he walked every morning at 10 o'clock, rain or shine. He would have discussions with his colleagues or study his collection of works by Lewis Carroll, which comprised every edition in every conceivable language.

He was a systematic man who liked order, regularity, and precision. His five children all had two first names with a total of 15 letters: Francis Falconer, Ethel Wordsworth (Venetia's mother), Nigel Cornwallis, Geoffrey Spencer, and Beatrice Gresley. The cook and the housekeeper served breakfast every morning at exactly five past eight. And as a true-blue Englishman, at the age of 80 he was still playing cricket regularly in the large garden behind his house.

G. Schilling, *The Hunt for Planet X*, DOI 10.1007/978-0-387-77805-1_5,
© Springer Science+Business Media, LLC 2009

Venetia Burney at 11 (Venetia Phair-Burney, courtesy Lowell Observatory)

Falconer Madan had an insatiable interest in every subject under the Sun and knew everyone who was anyone in Oxford personally. To little Venetia he was like a kindly old father.

Earlier that school year, Grandpa Madan had listened with interest to Venetia's stories about the nature walk that Miss Claxton had organized at the Parents' Union School. The walk through University Park had turned into a journey through the solar system, to give the pupils an idea of the great distances between the planets. The Sun was a circle on the blackboard, with a diameter of 60 centimeters. The Earth was a pea 200 steps away. To put the golf ball that represented Saturn in its place, the pupils had to walk nearly a kilometer. They had to forget about Uranus and Neptune that day, because they were too far away. Venetia's interest in the planets was aroused even more when she heard that their mysterious names all came from Roman mythology. Miss Claxton and her class read and re-read the book *The Age of Fable* by the nineteenth century author Thomas Bulfinch, and Venetia knew all there was to know about the kingdom of the gods.

The Pluto Breakfast

And then there was that breakfast, at five past eight on the morning of Friday, March 14, 1930. Grandma was not yet up – she always had breakfast in bed, at exactly 8.45 – and Venetia sat at the table with her mother and grandfather. Like every other morning, Falconer Madan was reading *The Times*, and he enthusiastically read aloud the report on the discovery of a ninth large planet in the solar system by American astronomers at Lowell Observatory. It was

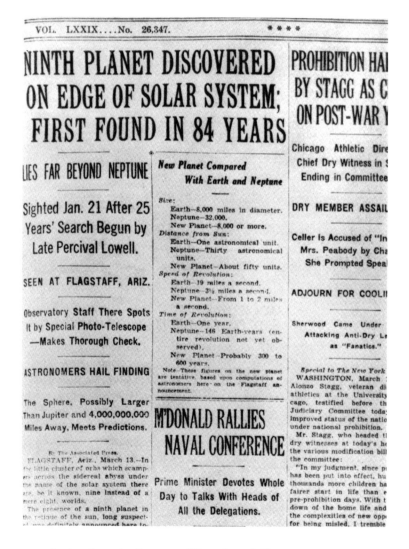

Newspaper clipping about the discovery of Pluto (Lowell Observatory)

smaller than Uranus, but probably larger than the Earth, said the paper, adding that it had not yet been given a name.

Falconer had a thing about names and celestial bodies. More than half a century earlier his older brother Henry had proposed the names Phobos and Deimos for the two small moons that had been discovered around Mars in 1877. But Henry had been dead for almost 30 years. So who would come up with an appropriate name this time? 'I wonder what they will call this new planet,' Madan said to Venetia and her mother.

Venetia was silent for a moment. She thought of the enormous distances in the outer regions of the solar system, the great expanse of dark nothingness, and the myths about the shadowy underworld. Then she said: 'They could call it Pluto.'

Pluto. It was of course a fantastic name. Madan himself had been thinking of Odin, the chief god of Norse mythology, but Pluto was much better. He had to inform the astronomers in Flagstaff about Venetia's idea as soon as possible.

That same morning, as he walked to the Bodleian Library, Madan slipped a note through the letter box of his friend Herbert Turner, who lived in nearby Blackhall Road. Turner had once been Astronomer Royal and, although he now concerned himself more with geology, he was still one of the leading astronomers in England. Perhaps he could pass on Venetia's suggestion to Vesto Slipher?

Herbert Turner (Royal Astronomical Society)

In the afternoon, on his way home, Madan dropped off another note to Turner, this time with a pound note enclosed. If Turner would be so kind as to send a telegram to Flagstaff, Madan was of course only too willing to cover the costs. What Madan did not know was that Turner was in London that day attending a meeting of the Royal Astronomical Society. And during the evening, the American discovery was of course the main topic of conversation. The astronomers came up with all kinds of creative names for the new planet, but Pluto was not one of them. When Turner returned home on Saturday afternoon and read the two notes, he sent a telegram to Flagstaff and informed Madan that 'Miss Venetia will get the best chance I can give her. [...] I scarcely think they will find a better name.' And he returned the pound note.

Andrew Crommelin, one of the most renowned astronomers of his time, was also enthusiastic about the name Pluto. In a letter to Turner on March 17, he described how it had been proposed at the end of the nineteenth century for the asteroid that was later called Eros, but had been rejected because it was too somber. But it was a perfect name for this new planet in the outer regions of the solar system. And didn't the mythological Pluto also have a helmet that could make him invisible? That was also very appropriate for a planet that had escaped attention for such a long time.

Cartoon Character

The proposal from Oxford also had – needless to say – competition from Cambridge, where a doctor called Archibald Garrod had also suggested the name Pluto. 'But,' Turner wrote to Madan, "I have told Sir Archibald that he is just a week behind 'a small girl', whose identity I will reveal if the name is adopted."

That was in the first half of April. Vesto Slipher and his colleagues were quite pleased with Venetia's proposal. Constance Lowell, the headstrong widow of the man who had started the hunt for Planet X, initially thought that Zeus was a good name for Percival's planet, then changed her mind and found Constance much more suitable. The astronomers in Flagstaff had been considering names like Minerva, Cronos, and Persephone, but everyone was very enthusiastic about Pluto. Not least because the first two letters of the name (which would also be the symbol of the new planet) were Percival Lowell's initials.

On May 1, 1930 the name Pluto was officially announced in a circular from Lowell Observatory. Falconer Madan, who had told his granddaughter nothing about his efforts to get her proposal accepted, was so pleased that he gave her five pounds to celebrate the good news – at that time an extraordinarily large gift for a child of 11. And he gave Miss Claxton a gramophone for her music lessons. The gramophone was promptly christened Pluto and was still in use 20 years later.

TY PAGES   PRICE FIVE CENTS

PLUTO CHOSEN AS NAME FOR PLANET 'X' DISCOVERED AT OBSERVATORY IN FLAGSTAFF

Nephew of Dr. Lowell Makes Announcement of Choice As Consistent With Names of Other Roman Gods Already Used

NEWS MADE PUBLIC AHEAD OF SCHEDULE

'Leaks' From Printing Office Where Bulletin Was Being Printed for Presentation to Two Astronomical Societies

FLAGSTAFF, Ariz., May 24—(AP)—Roger Lowell
Putnam, trustee of Lowell observatory and nephew of the late

Newspaper clipping about the naming of Pluto

The discovery of the new planet – and the first by an American! – was the talk of the day in the United States, and the name Pluto became very popular. And it was not because Pluto was already famous as Mickey Mouse's dog, because in May 1930 Mickey had not yet acquired a dog. The first time he had one was in the Disney cartoon *The Chain Gang*, which was released in September of that year, and it had not yet been given a name. A month and a half later, in *The Picnic*, the loose-limbed dog had been given the name Rover. But that name did not last long: in *The Moose Hunt*, which had its première on May 3, 1931, Rover had been renamed Pluto. According to the Walt Disney Studios there are no official documents to show that the cartoon dog had been named after the new planet, but no one doubts that Disney's films and comic strips have helped make Pluto the favorite planet of children.

Venetia Burney's role in giving a name to the new planet tended to be forgotten in the course of time. This is perhaps not so surprising since her own part in the affair was less important than that of her grandfather. For many years she did not give it much thought, nor did she maintain a great interest in astronomy. She studied and taught economics and, shortly after the end of the Second World War, she married the mathematician Maxwell Phair. And, no, the story of Pluto was not the first thing she told him. That only happened many years later, by coincidence.

At the end of the 1960s, when Maxwell had just retired, the couple took a trip around the United States, including a visit to Flagstaff, where they saw the telescope which had been used to discover Pluto. By then Clyde Tombaugh had been working for many years at the New Mexico State University in Las Cruces, and he and Venetia unfortunately never met. It had also been clear for a long time that Pluto was much smaller than what Tombaugh and Slipher had thought in 1930; it was certainly not larger than the Earth, as had first been assumed. And when a moon was discovered orbiting Pluto in 1978, which made it possible to determine the mass of the planet, it became obvious that Pluto really was a 'kid planet': it was not only smaller than all the other planets, but also quite a lot smaller than our own Moon. Poor Pluto was the runt of the solar system, too insignificant to take seriously. No wonder that young children were concerned about its fate.

Scrapbooks

In 1980, half a century after the discovery of Pluto, Tombaugh wrote the book *Out of the Darkness: The Planet Pluto*, together with the British author Patrick Moore. Venetia Phair-Burney, now 61, read the book a couple of years later. It was good to read about the discovery, of course, but why was her name not mentioned? Did Moore perhaps not know her story? Together with her husband she wrote a letter to the flamboyant author, broadcaster, and astronomy popularizer, and a few months later they paid him a visit at his house in Selsey, with Falconer Madan's scrapbooks under their arms. Moore promptly wrote a well documented article about how Pluto got its name for the American monthly *Sky & Telescope*, which was published in November 1984.

Venetia still has her grandfather's two scrapbooks, although she plans to donate them to the library at Lowell Observatory. They are a goldmine for historians of Pluto. Madan was exceedingly meticulous: he kept copies of all the letters he sent, an extensive journal, and newspaper clippings of all reports relating to Pluto. Without the scrapbooks, Venetia would probably not remember the events of the spring of 1930 as clearly as she does. After all, who remembers at such an advanced age what happened when they were 11 years old?

After Patrick Moore's article in *Sky & Telescope* more people started to show an interest in Venetia's story. The National Space Centre in Leicester invited her to the opening of its new planetarium and, although she could not attend, when she did visit the centre a few months later, she was welcomed as though she were the Queen herself. Later, too, her door was always open for journalists, radio interviewers, and television crews. And then, of course, there were the piles of letters from American schoolchildren wanting to know exactly how it had all happened and how you can write history at the age of 11.

A small asteroid discovered by amateur Japanese astronomers Seiji Ueda and Hiroshi Kaneda in 1987, and which had been provisionally designated as 1987 VB (Venetia Burney's initials), was later officially named (6235) Burney, so Venetia herself has now been immortalized in the solar system. But all the fame has not gone to her head. Most of her neighbors in Epsom have no idea that one of their number gave Pluto its name. After all, let's be honest, it is not something you can easily slip into the conversation unnoticed.

Venetia Burney has never actually seen 'her' planet. The kid planet is too faint to see with binoculars or a small amateur telescope. And it is doubtful whether she will still be around when the American space probe New Horizons arrives at Pluto, in the summer of 2015. But an experiment on board the probe is named after her: the Venetia Burney Student Dust Counter, developed by young students at the University of Colorado in Boulder. And there is little doubt that a striking crater or mountain top on Pluto will soon be named after an 11-year-old schoolgirl.

Chapter 6
A Strange and Wonderful Week

The dogs bark incessantly to the north of Townsend Winona Road in Flagstaff. This is the edge of town and the roads are sandy. The house on Golden Eagle Drive looks out on the cinder cones of the Sunset Crater National Monument, further to the north. The Sun has just gone down, and the summer heat hangs like a blanket over the dusty drive.

Jim Christy loves Arizona and Flagstaff. This is where Pluto was discovered, more than 75 years ago. What better place to live for the man who discovered Pluto's moon? Christy never wants to leave. He doesn't like to travel.

James Walter Christy is a big, sturdy man, with a penetrating but friendly expression, a firm handshake. He waits patiently, his lips resolutely closed, until he can tell the story of the discovery of Charon, a story he has told countless times. Then the words come out like a cascade. As though he is compensating for the extreme silence of his childhood. It doesn't worry him that he is sometimes difficult to understand. He knows he has a speech impediment, but he has no idea why, and no one ever mentions it.

It was June 1978, nearly 30 years ago. It was the strangest and most wonderful week of his life. A week in which the cosmos looked back at him for the first time. At Jim Christy, the quiet, old-fashioned observer from simple origins, with a bad memory and permanent self-doubt. He hasn't believed in God since he was 30, but you never lose your susceptibility to a near-religious experience.

Christy's father came to America from Poland with his parents as a small child. They settled in Milwaukee, on Lake Michigan. When he married in 1936 and became a naturalized American citizen, it seemed logical to change his complicated Polish name. Witold Chruszczynski became Walter Christy. But he never gave up his faith. His son Jim, born in 1938, was brought up as a strict Catholic and was initially determined to become a priest.

But things turned out differently. Jim contracted a kidney infection and doctors said he would be much better off living in a warm climate. So, in 1954, the family moved to Tucson, in the south of Arizona, where it was smoldering hot and the nights were pitch black. They lived a few streets away from the Steward Observatory. Walter and his son had once built a telescope together and little Jim had spent a lot of time in Milwaukee looking at his favorite star, Vega. So it was a logical step to study astronomy in Tucson.

G. Schilling, *The Hunt for Planet X*, DOI 10.1007/978-0-387-77805-1_6,
© Springer Science+Business Media, LLC 2009

Jim Christy at the StarScan plate measuring machine with which he discovered Pluto's moon
Charon (USNO, courtesy Jim Christy)

Then Christy had a stroke of luck. In October 1957, the Soviets surprised
the world by launching Sputnik 1, the first satellite. The United States
responded by setting up the Moonwatch program in great haste to observe
Sputnik as closely as possible. Christy's dedication and accuracy did not go
unnoticed and in September 1962, even before he had graduated, he was
given a job at the United States Naval Observatory in Flagstaff, where a
new telescope had been put into operation which could measure the positions of stars with great accuracy.

It was a quiet job, with many lonely nights at the telescope, sometimes 3
weeks or more at a time. Back then, Flagstaff was still a small, orderly
town. On Fridays, Christy and his colleagues would always have lunch with
the astronomers from neighboring Lowell Observatory. He became
acquainted with Vesto Slipher, learned about how Pluto had been discovered, and started a program at Mars Hill to make photographic observations of binary stars.

Missed Opportunity

It was there, in 1965, that he missed his first opportunity to discover Pluto's
moon. Lowell astronomer Otto Franz had taken two photographs of Pluto, on
which the point of light that was the planet did not appear to be perfectly round.

He asked Christy to take a look at them. Christy was pleased to oblige. He concluded that there was nothing abnormal about the photographs, other than that they were a little over-exposed. The idea that Pluto might have a moon simply did not occur to him. Pluto did not have a moon, everyone knew that. Planet expert Gerard Kuiper had searched for one with the world's largest telescope and found nothing.

Christy soon forgot Franz's photographs. He had other things on his mind. The Naval Observatory sent him back to Tucson in 1966 to complete his Masters degree. He was not all happy to be back at college. And that was not the only setback. His 4-year-old son fell out of a bunk bed and broke an elbow. His young daughter contracted tuberculosis. His wife was in hospital with a stomach ulcer. And shortly after he had resumed his studies, his 10-year younger brother was killed one night in a car crash. It was a bad year for Jim Christy – strange and terrible. It was around that time that he lost his faith in God.

And that was not the end of it. Christy's marriage ran into trouble a few years later and in 1972, while he and his wife were getting marriage counseling, he was suddenly and without explanation transferred to Washington D.C. It was a large, busy city and he had to commute an hour and a half from an apartment in Manassas, Virginia, to the headquarters of the United States Naval Observatory on Massachusetts Avenue.

It was a tedious form of astronomy that Christy practiced there. Using the StarScan, a professional plate measuring machine, he had to determine the exact positions of many thousands of stars, which were recorded on photographic plates sent to Flagstaff. It was a painstaking work which gave him headaches, and which always took place in a darkened room so that the enlarged image of a minute part of the night sky was clearly visible on the screen – along with the dust particles, hairs, and the occasional insect that find their way onto photographic plates, magnified to 30 times their normal size. Christy will never forget the startled expression on the face of a fire inspector who appeared in the darkened room through a little-used back door, unannounced and with much bravado. He turned on his heel and left again immediately when he saw a gigantic fly on the screen, apparently on the disturbing assumption that the US Navy was engaging in very top-secret experiments.

Photographs of the moons of Uranus and Neptune were measured at the laboratory, in preparation for the American Voyager project, which was to send unmanned space probes to the edge of the solar system. In the mid-1970s, Christy suggested also taking pictures of distant Pluto to look for moons. After all, the Naval Observatory had the best conceivable instruments to make the required observations. But his official proposal was rejected and, 2 years later, Christy had forgotten it completely. He was more occupied with his divorce and his marriage to the new love of his life Charlene in 1975 than with the solar system.

Omen

And then, in June 1978, came that strange and wonderful week. Jim and Char had bought a house in Berwyn Heights, just outside Washington D.C., and were due to move from their apartment on Monday, June 26. Jim was going to take the whole week off for the move and to get the house ready. For the first time in his life, he wanted to forget about astronomy completely for a while. That meant that he had to work extra hard the week before. So on Sunday, June 18, the conscientious Christy spent the whole night at the observatory in Washington working with the astrograph, a telescope that in effect worked like a huge telelens with a photographic plate in the back.

At sunrise on Monday morning, as always, he put the astrograph in its stand-by position, pointing straight up, and before he closed the dome, he took a look through the small finder telescope mounted on the side of the instrument. He had no particular reason for doing so but there, sparkling exactly at the crosshairs of the eyepiece, was Christy's favorite star, Vega. He had never realized before that its declination was practically identical to Washington's northern latitude so that it passed directly overhead. Is there a rational explanation for such a 'coincidence' or was it a good omen? Either way, it had a deep emotional impact on Christy: he had been looking at the stars for 20 years and now, for the first time, one seemed to be looking back. He had never felt such a strong bond with the cosmos.

On Tuesday morning, he had another strange experience. The StarScan could measure the positions of stars to an accuracy of a micrometer, but the

Jim and Charlene Christy in Arizona, around 1980 (Courtesy Jim Christy)

display always showed an extra number after the decimal point, which could be any random figure between zero and nine. To kill time, Christy tried to guess what that figure would be next time. To his amazement he guessed it right. Not once or twice, but three times and then four. It was starting to get creepy. When he got it right for the fifth time, he dared not go any further. The chances of predicting five random numbers in a row were one in a hundred thousand. Christy should have spent the day in Las Vegas rather than Washington. He would remember that experience for the rest of his life.

On Wednesday, June 21 – a few days earlier than planned – he had finished all the work he had to get done before taking leave. He asked his boss Bob Harrington if he had a small job to keep him busy for the rest of the week. Harrington pulled a couple of photographic plates out of the drawer which might otherwise never have been measured. They were the photos of Pluto which Christy had asked for a few years previously. They had not been exposed until earlier that spring. The astronomers in Flagstaff had seen that the images of Pluto on the plates were not perfectly round and had written 'IMAGE DEFECTIVE' in large letters on the plate covers, meaning that the images were unsuitable for serious astronomical work.

Normally, Christy would not have even bothered to look at the photographs. But he had nothing else to do and was happy to have something to keep him occupied. So, on Thursday, June 22, the plates went under the microscope and he realized that there was something strange about them. The images of Pluto were indeed slightly asymmetrical, as though the movement of the telescope had not been completely synchronized with the rotation of the stars. But the images of the stars themselves were perfect. So there had to be another explanation for the strange distortion of Pluto.

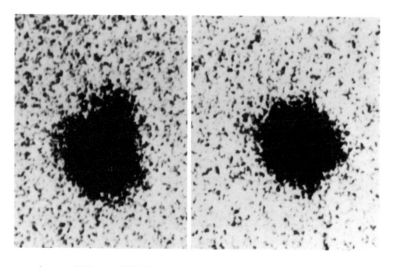

Discovery photos of Charon (USNO)

And what was even stranger – on one photograph the bulge was on the northern side, while a month later it was to the south. The word 'month' stuck in Christy's mind. Suddenly it was clear... 'Pluto has a moon,' he said aloud, although he was alone in the darkened room. He realized immediately that no one would ever believe him. After all, who was Jim Christy? He had never got his PhD, had few publications to his name, and now here he was going on about some crazy bulge on Pluto.

He wondered whether he should mention it at all. Imagine it was true. Then he could forget moving house the following week. Late on Thursday afternoon, he decided to tell Harrington anyway. 'Jim, you're crazy,' he said. Harrington was an exceptionally brilliant and sharp scientist – perhaps he was right. But Christy could not forget it. On Friday morning he took out a pile of old photographs of Pluto – including those made by Otto Franz in 1965 – and saw the bulge on them, too. It was not so difficult once you knew where to look. On a series of pictures from 1970, you could even see the little lump rotating around Pluto.

Later that morning, he had convinced Harrington, too. Pluto had a moon. And a relatively large one, at that. It was quite close to Pluto, and had an orbital period of just over 6 days, exactly the same as Pluto's rotation rate.

Everyone believed Jim Christy and took him seriously; his reputation was clearly better than he had ever thought. Within 2 weeks, the discovery had been confirmed by another laboratory, and on Friday, July 7, the International Astronomical Union published circular 3241 announcing it to the world.

Jim Christy (seated) and Bob Harrington (USNO, courtesy Stichting De Koepel)

Charon

But Christy's strange and wonderful week was not yet over. On the morning of Saturday, June 24, Christy and Charlene were driving to her parents' house. Char – as her family always called her – was of course very proud of her husband's discovery. Christy had been giving some thought to what the new moon was going to be called. He liked Oz, after the magical country in *The Wizard of Oz*. But while they were driving, he suddenly burst out: 'I could call it after you! How about Charon?' He pronounced it SHAR-on. The '-on' at the end made it sound genuinely scientific, like electron or neutron.

But during the weekend Christy's colleagues at the Naval Observatory had also been thinking about an appropriate name and had come up with Persephone. Persephone was the daughter of Zeus and Demeter, and had been kidnapped by Hades (the Greek name for Pluto), who made her queen of the underworld. Christy would never have come up with something like that himself – he knew almost nothing of mythology. And he had to admit it was a good name. His own proposal did not stand a chance.

And then came the strangest moment of his whole life. It was the night of Monday, June 26. Most of their possessions had been moved to the new house in Berwyn Heights on that warm summer's day, but the power had not yet been switched on. Jim and Char had gone to sleep exhausted, with a torch next to the bed. Around 1 o'clock, Christy suddenly woke up. If Pluto's moon was going to be called Persephone, he would have to tell Char the following day. He would not be able to keep his romantic promise. Or perhaps Charon had a chance for some other reason?

Against his better judgment, Jim got up, took the torch, and went to search for the box that contained his encyclopedia. When he found it, he took out the volume with 'C', sat down on another box, and with trembling hands started to leaf through it, reading in the light of the torch. And there it was, like a divine revelation: Charon was the ferryman who carried the dead across the River Styx to the underworld, Pluto's kingdom. Christy put the book back in the box, returned to bed, switched off the torch, and fell asleep.

The following day he had never felt so self-confident in his life. He told Harrington that the moon he had discovered would be called Charon, after the ferryman of the underworld. That the name would officially had to be pronounced CHA-ron didn't matter much to him – he could always stick to his own pronunciation. 'Many men promise their wives the moon,' Charlene Christy said, 'but mine delivered.'

After July 7, Jim Christy was suddenly a celebrity. His son flew in from Tucson and decided to stay and live in Washington instead of with his mother in Arizona. Young astronomers saw Christy as an example, a scientific hero. Slowly but surely it became clear to him that his accidental discovery had had a great impact on the careers of others.

In 1982 the Christy family moved to Tucson, which had now become a large city in itself, where Jim worked for 17 years in the physics department of the Hughes Missile Systems Company. But he continued to stay abreast of new findings on Pluto and Charon closely. And when he retired, he was finally able to return to his beloved Flagstaff where, in the thinly populated suburbs, it is still pitch black at night.

In the evening, he points to the silhouette of the ridge to the west of Flagstaff, where Lowell Observatory is located. High in the northeast twinkles Vega. Here it is never directly overhead, but that makes it no less special. And a little lower above the southwestern horizon, somewhere in the constellation Serpens, you should be able to see far-off Pluto. Since the discovery of Charon it has traveled a little more than 10% of its orbit around the Sun.

Christy puts his hands in his pockets, closes his lips firmly, and looks at the stars. Once, they looked back. But that was long ago. Tonight, only the dogs are barking. Without a doubt, there will be one among them called Pluto. Perhaps with a puppy. Strange, and wonderful.

Chapter 7
Fortunate Circumstances

The night of June 12, 2006 is crystal clear. And cold. Lake Tekapo, 300 meters below the University of Canterbury's Mount John Observatory, shines eerily in the silver-white light of the full moon. The stars of the Southern Cross twinkle above the southern horizon. Below the Moon to the right, in the constellation of Serpens, Pluto moves in front of a star. But the domes of the observatory remain shut. This rare event cannot be observed.

Astronomers from throughout the world have traveled to Australia and New Zealand to see this unique stellar occultation. Some of them have booked telescope time at Mount John Observatory. But the day before, the South Island of New Zealand was hit by the severest snowstorms in many years, and for miles around Lake Tekapo the roads are blocked and there is no power. After traveling so far, the astronomers lie fully clothed under thick blankets trying to keep themselves warm, while the observatory telescopes stand idly by.

Astronomy is an adventurous science. The cosmos sets the agenda; if you want to see rare celestial phenomena, you have to be in the right place at the right time. Astronomers go to Antarctica to witness a total solar eclipse, to Mongolia for a meteor shower or to New Zealand for a stellar occultation. But no matter how well you prepare your expeditions, the weather can always put an unexpected spanner in the works.

Marc Buie knows all about such disappointments. The Pluto expert from Flagstaff traveled with colleague Trina Ruhland to Carter Observatory in Wellington, on the southernmost point of North Island. On the evening of Monday, June 12, it was a crystal clear night, but during the night it got steadily cloudier. And although they were able to take some measurements through the cloud cover, they were of little scientific value. It was disappointing of course, but you get used to that in their line of work.

Buie's team-mate Eliot Young from the Southwest Research Institute in Boulder, Colorado, was at a farm in Tasmania, together with Jeff Register. Around 4.30 in the morning, when the stellar occultation occurred, the sky was clear but the mobile observing station that Young and Register had taken with them was not frost-proof. Condensation in the telescope and camera ruined their measurements, and there was a short circuit in the electronic timer.

G. Schilling, *The Hunt for Planet X*, DOI 10.1007/978-0-387-77805-1_7,
© Springer Science+Business Media, LLC 2009

Marc Buie (standing) and Trina Ruhland at Carter Observatory (Courtesy Marc Buie)

Lesley Young, Eliot's sister and colleague, had set up an identical mobile station, together with Cathy Olkin and Ross Dickie, near Omakau, but when it started to get misty about half way through the evening, they decided to pack everything up and move to Wanaka, an hour and half to the northwest. Unfortunately, later that night it got cloudy there too and even started to snow again.

Richard French of Wellesley College in Massachusetts had better luck. French and his colleague Kevin Shoemaker used the 3.9-meter Anglo-Australian Telescope on Siding Spring Mountain in New South Wales, Australia. According to the predictions, the stellar occultation would not be visible from there, but French and Shoemaker decided to give it a try anyway. When they sent their observations through to Buie that same night by email, Siding Spring proved to have produced the best results.

Atmosphere

A stellar occultation is in effect a sort of solar eclipse. With a total solar eclipse, the Moon passes in front of the Sun. To observers in the shadow of the Moon, the light of the Sun is blocked. The distant dwarf planet Pluto can, of course, never pass in front of the Sun – that would mean it has to pass between the Earth and the Sun. But like every other planet, Pluto does occasionally pass in front of a star. And since stars are in effect distant suns, you could call it a kind of solar eclipse. On the night of June 12, 2006, the astronomers who had traveled all the way to Australia and New Zealand were actually in the 'shadow' of Pluto – the shadow cast by the extremely weak star UCAC 2603 9859, to be exact.

The first serious attempt to observe a stellar occultation by Pluto was made in 1965. Unfortunately, at the time the orbit of the dwarf planet was not known accurately enough to predict exactly where the event would be visible. This continued to be a problem in the years that followed. But in 1980, South African astronomer Alistair Walker finally struck lucky. Instead of Pluto, however, he saw its moon Charon pass in front of the star. From the duration of the eclipse, Walker was able to calculate that the moon had a diameter of a little less than 1,200 kilometers.

A stellar occultation by Pluto itself was not actually observed until Thursday, June 9, 1988. By coincidence, it was only visible then too from Australia, New Zealand, and the South Pacific Ocean. A shame for Marc Buie, who was then working at Mauna Kea Observatory on Hawaii, from where the eclipse could not be seen at all. Fortunately many other astronomers had decided to travel to the best locations. One team had even decided to observe the occultation from a converted plane, to make sure that the clouds would not spoil the party. And all their efforts were not in vain. On the contrary: it was discovered that Pluto has an atmosphere.

From the Earth, a star is a dimensionless point of light. The same is true for Pluto – even with a large telescope, it is impossible to discern the characteristic disk-shape indicating a planet. Just before a stellar occultation the two faint points of light – the star and the dwarf planet – are very close together in the night sky. They can no longer be seen individually and a telescope sees the light they radiate to merge together into a single source. During the eclipse, which lasts less than 2 minutes, the star is invisible and only the light from Pluto remains.

If Pluto were just a bare ball of rock or ice, the star's light would go out suddenly. But on June 9, 1988 that happened only very slowly, taking several tens of seconds. It was as though Pluto had a very thick atmosphere, which absorbed steadily more of the starlight. But in fact, the explanation is quite different. The atmosphere is actually very thin and bent some of the star's light even after it had already disappeared behind the planet's edge. In this way, light appeared in Pluto's shadow and the star seemed to fade slowly.

The exact way in which the starlight is bent depends, among other things, on the air pressure in Pluto's atmosphere. Measurements of the brightness variation of the star showed that the pressure was only a few microbars – less than a hundred thousandth of the air pressure here on Earth. Strangely enough, 14 years later, in 2002, observations by various teams of *two* stellar occultations by Pluto (on July 20 and August 21) showed that the air pressure had roughly doubled. This was evidenced by the star fading in a slightly different way.

Expedition to Chile in July 2002 to observe a stellar occultation by Pluto (Courtesy Marc Buie)

That was surprising, because in 1988 Pluto was 4.46 billion kilometers from the Sun (29.8 times as far as the Earth) while in 2002 it was 4.58 billion kilometers (30.6 astronomical units). Astronomers had expected the atmosphere to actually be *thinner* because of the greater distance from the Sun and the corresponding lower temperature. But even in June 2006, when the distance from Pluto to the Sun had increased further to 4.67 billion kilometers (31.2 AU), the air pressure was still just as high as in 2002.

Mr Pluto

Marc Buie from Lowell Observatory in Flagstaff can't get enough of these inexplicable Pluto mysteries. The very fact that so little is known about the distant dwarf planet means that every measurement brings yet another surprise. Buie is known as 'Mr Pluto,' and not without reason. His whole life is devoted to the small, cold dwarf planet. If he could go there tomorrow, he would jump at the chance. And he would not be deterred by the thought that the trip would take so many years.

Even before Neil Armstrong was the first man to set foot on the Moon, Marc knew that he wanted to be an astronaut. You had to be trained as a pilot, so together with a friend from his hometown of Baton Rouge, Louisiana, he wrote letters to airline companies to find out how to go about it. But their boyhood dreams were all shattered when the bespectacled high-flyers heard that you needed excellent eyesight. So much for a career as an astronaut.

But this did nothing to dampen Marc's interest in everything extraterrestrial. He devoured science fiction books, scoured the heavens with a small telescope, and refused to be discouraged by a bad astronomy teacher at Louisiana State University. And so it happened that in October 1982 Buie was a promising Master's student in planetary research at the Lunar and Planetary Laboratory at the University of Arizona. And that he found himself facing an oral examination to defend two proposals for his final thesis.

But a week before the exam, when Buie should have been studying the vertical structure of the atmosphere of Venus and the surface properties of Jupiter's moon Io, he came across an article by his supervisor Uwe Fink which argued that Pluto must have a very thin atmosphere. This was the only explanation for the frozen methane on the surface, which had been discovered in 1975.

What kind of world was this? Small, dark, and cold, with a temperature a few tens of degrees above absolute zero, crystals of methane ice, a thin atmosphere, and a large moon... Marc Buie was hooked. He had lost his heart to Pluto. At the last moment he changed the subject of his thesis, conducted observations of Pluto with the 1.5-meter Kuiper Telescope at Steward Observatory in Arizona, and gained his PhD in 1984 for his spectrophotometric research into the remote ice planet.

Pure Luck

Buie couldn't have timed it better. Early in 1985 a rare series of mutual eclipses and occultations started between Pluto and its large moon Charon. Charon orbits

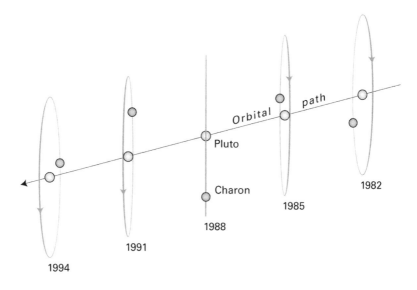

Between 1985 and 1990, Pluto and Charon displayed mutual eclipses and occultations, as seen from the Earth (Wil Tirion)

Pluto in a little over 6 days, and from the Earth we see this orbit from various angles. Generally, it is slightly tilted to a certain degree, but once in a while, the orbit can be seen exactly edge-on. Pluto takes 248 years to revolve around the Sun, and twice in that time there is a period of a few years when Pluto, Charon, and the Earth can be exactly aligned.

The first time astronomers observed this was between 1985 and 1990. It was pure luck of course. If Charon had not been discovered in 1978 but 13 years later, the rare series of celestial phenomena would have already happened and Pluto researchers would have had to wait until the start of the twenty-second century.

But now the eclipses and occultations offered a fantastic opportunity to learn a lot more about Pluto and Charon. The idea is simple. If you know the sizes of the two bodies, the distance between them, and the angle of Charon's orbit, you can work out exactly when Charon's shadow will fall on Pluto, when Charon itself passes in front of Pluto, when it passes through Pluto's shadow, and when it is hidden behind the planet. And of course, it all works the other way around too: if you measure the eclipses and occultations very precisely, you can learn a lot about the two bodies.

The only thing you need for this is accurate brightness measurements. With most telescopes on Earth, Charon cannot be seen as a separate body, because it is far too close to Pluto. A sensitive photometer therefore always registers the combined light from both. But if there is an eclipse or occultation, the combined brightness will decrease. If you measure that brightness variation extremely accurately, it is even possible to learn more about the location of dark and light areas on Pluto and Charon. And if you record the brightness variations with a spectroscope, you can determine the composition of their surfaces.

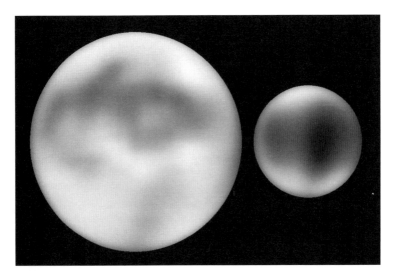

Crude maps of Pluto and Charon, as derived from their mutual eclipses and occultations (Adapted from images by Marc Buie/Lowell Observatory)

Learning about two worlds by studying a small, variable point of light in the night sky – that is astronomy at its best. And it's where Marc Buie is in his element. In close cooperation with David Tholen at the University of Hawaii – where Buie was a postdoc in the mid-1980s – he dedicated himself heart and soul to observing the mutual eclipses and occultations of Pluto and Charon. And when a stellar occultation by Pluto was observed for the first time on June 9, 1988, and the existence of an atmosphere around the dwarf planet was confirmed, it was time to celebrate. Ten years previously, before the discovery of Charon, almost nothing was known of Pluto; now it seemed that the small, icy double planet held no more secrets.

Odd Man Out

So what kind of a world is Pluto? It is certainly very different from all the other planets that have so far been found in the solar system. For a start it has a very strange orbit around the Sun. The orbits of the other planets are roughly circular, but Pluto's is an elongated ellipse. At one point it is 4.43 billion kilometers from the Sun (less than 30 AU and closer than the planet Neptune!), while half an orbit and 124 years later it is 7.38 billion kilometers away (nearly 50 AU). And while the Earth and the other planets revolve around the Sun in more or less the same plane, Pluto has an orbital inclination of 17 degrees, the equivalent of a gradient of 30%.

Pluto not only has an eccentric orbit, the planet itself is completely out of kilter. Like Uranus it lies 'on its side,' with its axis almost in level with the orbital plane rather than perpendicular to it. And it is small. Much smaller than Percival Lowell expected, smaller than Clyde Tombaugh had feared it would be. Pluto is less than half the size of Mercury and is even quite a bit smaller than our Moon. It is only about 2,300 kilometers in diameter, roughly the distance between New York and Houston.

The orbital motion of Charon already suggested that Pluto is a planetary lightweight – 500 times less massive than the Earth and almost 30 times lighter than Mercury. If you combine the diameter and the mass, it is easy to calculate Pluto's density: a little more than 2 grams per cubic centimeter. That means that the dwarf planet cannot be made of stone and iron, but must consist largely of ice.

A manned expedition to the ice world is not on the cards for the foreseeable future, but Marc Buie can very easily imagine what it must be like to walk around on Pluto: with less than 1% of your weight on Earth because of the very low gravity, at temperatures of 230 degrees below zero, in the twilight because the Sun is nothing more than a dazzling star in the black sky, across snowfields of methane ice and transparent crystals of frozen nitrogen, and with a gigantic moon hanging overhead – at least, if you are on the right side of the planet.

Artist impression of Pluto and Charon (Gemini Observatory/AURA)

Charon is about half the size of Pluto and revolves around it at a distance of less than 20,000 kilometers. The two bodies always show each other the same face, as though they are connected by an invisible rod. Pluto is invisible from the 'back' of Charon and anyone living on the wrong side of Pluto will not even know of Charon's existence. On Pluto, every 6 days, 9 hours, and 18 minutes, you would see the Sun rise and set again, the stars rotate around a point halfway between the stars Spica and Antares, and Charon passes through all of its phases.

It is a magical world. Small, distant, dark, and cold – an icy mini-planet in a peculiar orbit. And it is not a world that you will soon tire of. Certainly not, if you are Mr Pluto. Riccardo Giacconi, who was the director of the Space Telescope Science Institute in Baltimore, once asked Marc Buie what *else* he was intending to study, besides Pluto. But the young astronomer told him there is no time for anything else. You never tire of Pluto.

Just back from the occultation expedition to New Zealand, exhausted by lack of sleep and jetlag, Buie still talks again enthusiastically and expansively about the unexpected behavior of Pluto's atmosphere. The results show that it simply refuses to get thinner, even though the dwarf planet's elliptical orbit takes it further away from the Sun and the methane and nitrogen molecules must freeze solid on the surface sooner or later.

What will the next stellar occultation reveal? We will soon know: Pluto is moving slowly but surely closer to the Milky Way in the night sky and will encounter many more faint stars than in past decades. Will the volatile atmosphere have collapsed onto the surface? Or will there still be some of it left over in July 2015, when the space probe New Horizons arrives at Pluto? We have to wait and see. With Pluto you never know. It is full of surprises.

Chapter 8
Nix and Hydra

Max Mutchler's son Sawyer is sitting on his father's lap. It is time for him to go to bed, but first Max has to read him another story. From one of Sawyer's favorite books, about the planets of the solar system. Max leafs through the book quickly until he finds the page about Pluto, the planet Sawyer likes best.

'Look, it says it right here: Pluto has one moon, just like the Earth.'

Sawyer starts to beam and looks at his father full of pride.

'But is that right, what the book says?'

'No!' cries Mutchler junior.

'So how many moons does Pluto have?'

'Three!'

'And who discovered the other two?'

'Daddy!'

'But it's still a secret, ok? You mustn't tell anyone!'

Another hug and Sawyer is tucked up in bed. No, he won't tell anyone, but of course his daddy is the best astronomer in the whole world.

Max Mutchler was still a child himself when he became interested in the solar system. His grade four teacher had a copy of the book *The Search for Planet X* which had left a deep impression on little Max, especially the wonderful drawings. He now has a copy on his own bookshelf, bought recently from Amazon for $2.25. Published in 1962 by Basic Books, as part of the *Great Mysteries of Science* series, with text by Tony Simon and illustrations by Ed Malsberg. And with wonderful, inspiring passages. Like, for example: 'Within your own lifetime, you may yet see pictures of planet number 10 and have the fun of debating what to name it.'

Mutchler did not discover a tenth planet, but he *did* discover two small moons around the ninth. And if it ever took a long time to decide on a name for something, it was these two small frozen lumps of rock.

It all started in 2003. NASA had just given the go-ahead for the New Horizons project, the first unmanned space probe to Pluto. John Spencer of the Southwest Research Institute in Boulder, Colorado, submitted an observing proposal on behalf of a large group of Pluto researchers to use the Hubble Space Telescope to search for as yet unknown moons around Pluto. Hubble is of course excellently equipped for such a search, but the Space Telescope Science Institute in Baltimore receives so many proposals that it always has to apply a

G. Schilling, *The Hunt for Planet X*, DOI 10.1007/978-0-387-77805-1_8,
© Springer Science+Business Media, LLC 2009

Max and Sawyer Mutchler, sharing the secret of Pluto's second and third moons (Courtesy Max Mutchler)

strict selection procedure. And Spencer's proposal was rejected, as was a second proposal submitted early in 2004 by Spencer's colleague Hal Weaver of Johns Hopkins University, also in Baltimore.

Hubble Photos

Of course, there had already been searches for new moons around Pluto, with large ground-based telescopes. New Horizons' principal investigator Alan Stern had initiated the search at the end of the 1980s. And Brett Gladman and Philip Nicholson studied Pluto in 1999 with the 5-meter Hale Telescope on Palomar Mountain. But none of these searches bore fruit. It would be fantastic to make a new attempt with the Hubble Space Telescope, but that seemed impossible.

On September 28, 2004, however, Weaver and his colleagues received an unexpected message from the Space Telescope Science Institute. One of the science instruments on board Hubble, the Space Telescope Imaging Spectrograph, had stopped operating on August 3 as a result of an electronic fault. Since many of the planned measurements could now not take place, and to make the best use of the telescope time that had become available, some of the rejected proposals were being given a second chance. In May 2005, Hubble would therefore start searching for new moons around Pluto. One man's loss is another man's gain.

Max Mutchler knew nothing of all this. He had been working at the Space Telescope Science Institute since Hubble was launched in 1990, in more recent years as an instrument analyst on the Advanced Camera for Surveys, the extremely sensitive on-board camera. He is a real magician when it comes to 'scientific photoshopping' – cleaning up the rough images produced by the camera so that all undesirable noise is removed and every tiny detail is made visible.

On Monday, June 13, 2005 Hal Weaver brought Mutchler the electronic images of Pluto taken by Hubble on May 15 and 18. Weaver had not had time to examine the images in detail himself, but wondered whether Max could take a look? He told him to unleash all his magical skills to improve them.

Max didn't get around to it immediately. On Tuesday, June 14, his daughter Sierra and son Sawyer started their summer vacation and Max stayed home with the family. There didn't seem to be any great hurry. Surely Weaver's colleagues had also studied the images? Max was under the impression that they just wanted him to give a kind of second opinion. But no one had actually found the time to analyze the Hubble images, largely because team leader Weaver was much too occupied with the space probe Deep Impact, which was due to arrive at comet Tempel 1 on July 4.

Mutchler did not therefore call up the images on his computer monitor until Wednesday, June 15. His practiced eye almost immediately saw something interesting: a minute point of light just outside the overexposed images of Pluto and Charon. And, a little further away, another one. To record extremely small and faint satellites, the Hubble photos had been exposed for a relatively long time: 8 minutes. In that time, Pluto had moved very slightly in the sky, at approximately the speed of a snail at 1,500 kilometers distance. Any satellites it may have would of course move along with it and their faint images would therefore be spread out over a few pixels. To prevent that, Hubble was programmed so that it stayed focused on Pluto while the photos were exposed. Faint moons would then remain point-like (in those 8 minutes they would hardly move along their orbits) while all stars in the background would appear as short stripes of light.

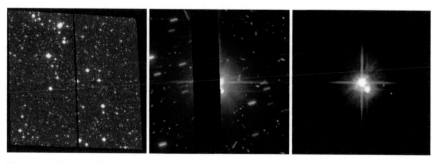

Discovery images of the new satellites of Pluto. *Left*: Full Hubble Space Telescope frame. *Middle*: Zoom-in on Pluto, surrounded by many bad pixels. *Right*: Official discovery release photo (Courtesy Max Mutchler/NASA, ESA, H. Weaver (JHU/APL), A. Stern (SwRI), and the HST Pluto Companion Search Team)

The spots that Mutchler saw were not extended, and could therefore not be background stars. And they were not as sharp as the many 'bad pixels' on the Hubble photos – points of light that occur when the CCD chip in the camera is hit by high-energy cosmic ray particles. And the two specks were also visible on the photos taken on May 18, but then in a slightly different position with respect to Pluto. If you then assumed that they were small satellites revolving around Pluto in the same orbital plane as Charon, you could calculate their distances and orbital periods, where they would be in the future and where they must have been in the past. Could it be that easy to discover new planetary satellites?

Double Discovery

Mutchler did not of course take any chances. He applied all his knowledge and skills to the Hubble photos until he finally ended up with clear, clean images. But the points of light were still there. That evening, as he was about to go home to bed after another long day, Max sent a quick email message from Baltimore to Ted Biro, an old friend from his college days who now lived in Minneapolis. His last sentence was: 'I may have just found another moon around Pluto.... *zzzzzzzzzz*.'

On Thursday morning, Mutchler mailed Weaver to tell him about the find, but he did not reply. Unbelievable as it may sound, Max also forgot about Pluto and its moons for a while. All kinds of other projects demanded his attention, like the Deep Impact mission and a newly discovered supernova. And in between working, he also took a vacation with his wife and children in Montreal, Canada.

But on the other side of the US, in Boulder, someone else was also looking for satellites orbiting Pluto. Alan Stern of the Southwest Research Institute had just taken on a new postdoc – Andrew Steffl – and thought it would be a good idea to let him gain a bit of practical experience by analyzing Hubble images. Stern had planned and prepared the Hubble Pluto observations together with Weaver, but because Weaver was officially the team leader he had to check with him first. Weaver thought it was fine, of course, and Steffl started working on the images at the end of July. Weaver did not mention the small satellites that had already been found on the Hubble images. He probably thought that there couldn't be any harm in getting independent confirmation.

Andrew Steffl had little experience and tackled the job very thoroughly. He did not find the two satellites until Wednesday, August 17, 9 weeks after Mutchler had discovered them. That evening, under the shower, he suddenly had that unique feeling of discovery: Pluto has two extra moons and I am the only person in the world who knows about them. The next day he reported his find to Stern, who was of course very proud of his new postdoc. Together they conducted a series of checks to be absolutely sure that the points of light were not errors on the photos, and on August 24 they called Weaver to tell him that they had discovered two new satellites around Pluto.

It was a strange call. The festive mood in Boulder was of course dampened by the news that the moons had already been found in Baltimore. Unfortunately, only one person can be first and Max Mutchler would eventually go down in history as the man who discovered the two small satellites. Mutchler also finally received a reply from Weaver, who told him that he was 'probably right' about his discovery.

Of course, new photographs had to be taken to provide absolute certainty. Preferably, they should be taken by the Hubble Space Telescope, and if possible without having to go through the usual prolonged procedures of official observing proposals and committee meetings. Fortunately the director of the Space Telescope Science Institute has a certain amount of 'own' time on Hubble for such special cases. So, on August 30, Weaver and Stern submitted a Director's Discretionary Time Proposal.

In the meantime they agreed with Mutchler and Steffl that the news must not be leaked to the outside world. This was an important discovery that needed further confirmation and would then be announced in a wave of publicity. Mutchler hoped that his friend Ted would not pass on the last sentence of his message to a journalist. It was no problem that he had talked about it at home; his wife Julie would not tell anyone and he knew that Sawyer could keep their secret, too.

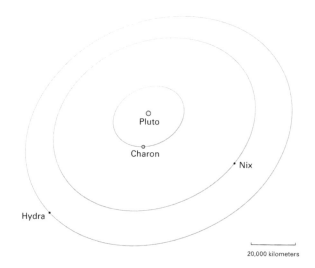

Orbits of the moons of Pluto (Wil Tirion)

Confirmation

But finding confirmation proved more difficult than they had expected. On August 29, two of the four still operational gyroscopes aboard the Hubble Telescope were deactivated as a precautionary measure. These sensitive instruments, which regulated the exact position of the telescope, displayed a sudden tendency to give up the ghost. Scientific observations were still possible with two gyros, but no longer in every corner of the night sky. Pluto would not come into range again until February 2006.

So what next? Together with Bill Merline of the Southwest Research Institute, Weaver and Stern contacted the directors of the largest ground-based telescopes, like the Keck Telescope and the Gemini North Telescope on Mauna Kea, Hawaii, and the European Very Large Telescope in Chile. They had better luck there. On Friday, September 9, the 10-meter Keck Telescope, the largest in the world, would take photographs to see if the two new moons were indeed visible in the predicted position. And later in the month, the Gemini North and the Very Large Telescope would also make observations of Pluto.

In the meantime, Max Mutchler was starting to get nervous. He knew that Mike Brown of the California Institute of Technology (Caltech) in Pasadena would be using the Keck Telescope a day later, on September 10. Early in 2005, Brown had discovered a large icy object well beyond Pluto's orbit (see Chapter 22) and wondered if this 'tenth planet' also had a moon. Would the Keck technicians keep quiet about the new Pluto moons? If they said anything about the photographs made the previous day, Brown might also try to observe the Pluto moons and it was not inconceivable that he might then announce the discovery himself.

Fortunately Mutchler's fears were ungrounded. But it also very quickly became clear that the big telescopes had been unable to detect Pluto's small, faint satellites. They were drowned out by the light of Pluto itself, which was a hundred thousand times brighter and had a much more disruptive effect on ground-based observations than on Hubble because of the influence of the Earth's atmosphere. More than 3 months after their discovery, the existence of the two small satellites had not yet been confirmed independently.

At the beginning of October, the whole affair started to become even more painful. During his observations with the Keck Telescope on September 10, Mike Brown had discovered that his 'tenth planet' did indeed have a relatively large moon. Caltech announced the discovery in a press release on September 30. On the basis of a single grainy image, and with no independent confirmation. So why couldn't Mutchler's discovery, which was based on far better quality photographs taken by Hubble on two different dates, also not be made public?

But Weaver and Stern didn't want to take the slightest risk. They didn't feel certain enough until October 24, when they got confirmation from Pluto researchers Marc Buie and Eliot Young. Stern had told them about the find and they had taken another look at their old Hubble photographs of Pluto from 2002. Because they now knew what they were looking for, they found the two new moons in the exact position they had expected them to be. A week later, on Monday, October 31, 2005, the Space Telescope Science Institute issued a press release and Sawyer Mutchler could finally tell his secret to his school friends. Pluto had two new moons, provisionally named S/2005 P1 and S/2005 P2 (the

S stands for satellite and the P for Pluto) and both around a 100–150 kilometers across. They were 64,700 and 49,400 kilometers from Pluto, and had orbital periods of just about 40 and a little over 25 days.

Names

Alan Stern had insisted that the news be announced in October. On November 1 and 2, there was a meeting planned at the Kennedy Space Center in Florida for all scientists involved in the New Horizons project and it would of course be an excellent opportunity to discuss the new discovery. New Horizons was to be launched early in 2006 and now had four objects to study when it passed Pluto in 2015 instead of two.

At Stern's request, Max Mutchler also went to Florida, where he met Andrew Steffl for the first time in person. Max knew exactly how Andrew must be feeling. He himself had once discovered a supernova in a remote galaxy only to find out later that it had already been discovered by someone else. So he could understand and empathize with Andrew – and admire him: if you can make such a discovery only 3 months after gaining your doctorate, you'll go far.

The weeks that followed meant a lot of hard work preparing scientific publications on the new Pluto moons but, of course, a lot of thought was also given to finding suitable names. Keith Noll, an astronomer at the Space Telescope Science Institute, had already spoken to Mutchler about possible names for the moons in September. Because of Steffl's 'shadow discovery,' they joked about calling them Baltimore and Boulder, but that was never a serious proposal.

Pluto satellite discoverers. *From left to right*: Andrew Steffl, Jim Christy, and Max Mutchler (Courtesy Andrew Steffl)

Mutchler knew about Jim Christy, who had in effect named the large moon Charon after his wife, but he felt that suggesting Julie, or Sierra and Sawyer, was going a little far. In any case, the moons had to be given appropriate names from Greek mythology. In other words, it was time for some serious brainstorming. All the team members were asked to submit suggestions and the most popular names would be selected through a series of votes.

Andrew Steffl liked the sound of Persephone and Cerberus. Persephone was the queen of the underworld and Cerberus the three-headed dog that guarded the gates of Hades. But those two names had already been given to asteroids. Max Mutchler thought Aether and Hemera would be good names. Firstly because they were the son and daughter of Erebus and Nyx, the god of darkness and the goddess of the night. But especially because the initials A and H also stood for Alan and Hal, who had led the search for the two satellites.

The names Obolus and Danake were also considered – in Greek mythology these were the coins the dead had to pay the ferryman Charon to cross the River Styx. But eventually, the team decided on Nyx (for the inner moon) and Hydra. Night goddess Nyx was not only the mother of Aether and Hemera, but also of Charon. Hydra was the nine-headed monster defeated by Heracles. The nine heads fitted in neatly with Pluto's status as the ninth planet, while the letters N and H referred to the space probe New Horizons.

In February 2006, when the New Horizons probe was well on its way, the Hubble Telescope finally took new photographs of the two small Pluto moons. They would be available at the Space Telescope Science Institute on the 16th of the month, and Mutchler had invited Weaver to come and take the first look at them. Imagine that there turned out to be a third moon on the photos – the team leader would at least have the honor of being the first to see it. But the images were ready a day in advance and Mutchler could not contain his curiosity. To his great relief, the photographs showed both moons exactly in the predicted positions. The press release about the new Hubble observations was issued almost at the same time as the scientific publication of the discovery, on February 23 in the British weekly magazine *Nature*.

Quibbling

Now that the original discovery had been confirmed so convincingly it was time to submit the proposed names officially to the Outer Solar System Nomenclature task group of the International Astronomical Union. And it soon became clear that the race was far from over. First of all, there was apparently already an asteroid called Nyx. And Hydra was not only the name of a constellation, but also of the German prostitutes' union, which led some of the members of the working group to argue that it wasn't a suitable name for a planetary satellite.

The discussion also rekindled an old dispute. In the spring of 2006 the debate on Pluto's true nature became more heated than ever. Alan Stern had always been firmly in favor of categorizing Pluto as a planet, but a lot of astronomers felt that the little odd man out did not merit such an elevated status. And many of them found it unacceptable that Stern and his colleagues now wanted to call one of the small moons of the 'ninth planet' after a nine-headed monster.

And so the members of the group came up with their own creative suggestions. What about Cerberus and Typhon, whose first letters formed the initials of Clyde Tombaugh, who discovered Pluto? Or Typhon and Ladon which, together with Pluto and Charon, formed the letters P, L, C, and T, the initials of Percival Lowell and Clyde Tombaugh?

The quibbling continued by email for several weeks. Never had there been such a fierce debate about the names of small planetary satellites. Obviously professional astronomers are very sensitive about anything to do with the distant dwarf planet Pluto. Finally, after several rounds of voting, the names proposed by the team that had discovered the moons were adopted, with Nyx now spelt the Egyptian way (Nix) to avoid confusion with the asteroid. The names of the two new moons were officially announced on Tuesday, June 20, a little more than a year after Max Mutchler had discovered them.

After the discovery of Nix and Hydra, Pluto became even more of an odd man out. Even though some people see the small body in the outer regions of the solar system as a full-fledged planet, it is completely different from the other eight. They fall into two clear categories: the relatively small 'earthlike' planets (in addition to the Earth: Mercury, Venus, and Mars) which consist of rock and metal, and the four giant planets (Jupiter, Saturn, Uranus, and Neptune) which comprise mainly light elements such as hydrogen and helium. Pluto is clearly not a gas giant. But it is also much smaller than the four terrestrial planets, comprises more than 50% ice, and now proves to have a complicated system of moons – one large one and two small ones, as though the moons of the Earth and Mars were all thrown together in one heap.

Max Mutchler remains stoical about the whole affair. He may go down in history as the man who discovered Nix and Hydra, and his son Sawyer may rightly be proud of him, but actually it was just a matter of chance. He was the right man in the right place at the right time, with the right photographs in his hands. It was nice of course to think back to that children's book by Tony Simon and Ed Malsberg and the words that had inspired him so long ago: 'Within your own lifetime, you may yet see pictures of planet number 10.' Or of two small moons circling the ninth – that's also something to tell your kids about.

Chapter 9
The Unauthorized Planet

Charles Kowal has found his place. Surrounded by nature, far away from the big city, with a magnificent view of Mount St. Helens and Mount Rainier. A beautiful house on Mayfield Lake in Cinebar, a small village in the state of Washington. Peace and quiet. And a lot of space. But Kowal does not live there alone. His self-built robots keep him company. He programs them from his computer and tracks their movements on television, where the images recorded by their miniature color cameras appear on the screen. His dream is to build a robot that will pour him a beer and will really care for him. But it's a dream that is a long way from being fulfilled.

Charles Kowal in Glacier National Park, Montana, in 1999 (Courtesy Charles Kowal)

Kowal is used to looking after himself. And to being alone with nature. Under the stars. Backpacking in the Rocky Mountains. Unexpected encounters with marmots. Or spending nights alone in a cold observatory. And speaking of dreams, he knew since he was a small boy that he wanted to be an astronomer. He could think of nothing better than working with the largest telescopes in the world, eye to eye with the universe.

That was half way through the last century in Buffalo, in New York State, where the night sky could still be seen in all its glory, especially during the bitterly cold winters. Charlie read *Sky & Telescope*, was a member of an astronomy club and scoured the skies at night with his amateur telescope. He was never happier than when he was observing the cosmos. Other worlds, faraway from the Earth.

Of course it was unusual for a 16-year-old schoolboy to move from the cold East to the sunny West coast, leave his parents' home, and study astronomy at the University of Southern California in Los Angeles. But it was certainly close to the action, because some of the world's largest and best telescopes were to be found on Mount Wilson and Palomar Mountain, where great astronomers like Edward Emerson Barnard and Milton Humason had worked. And didn't every teenager at that time dream of moving to California? Does it matter what the main reason was for going?

And Kowal was impatient. The sooner he could spend his time looking through the eyepiece of a telescope the better. After following a 4-year Bachelor's program, he wrote open application letters to four observatories. In the fall of 1961, while not yet 21, he started work at Mount Wilson, to the north of Pasadena. He hardly gave himself the time to complete his studies properly: he was not awarded his degree until 2 years later, after he had finished his final paper.

Painstaking Work

Mount Wilson Observatory had its headquarters in Pasadena, on the campus of the California Institute of Technology, but Kowal preferred to spend his time at the observatory itself, high in the San Gabriel Mountains where, between the tall fir trees, you had a spectacular view of Los Angeles. Or, even better but further away, on Palomar Mountain, to the north of San Diego, where the gigantic 5-meter Hale Telescope had been in operation since 1948. Both observatories fell under the same administration and, in the early 1960s, Pasadena was the beating heart of the astronomical world, where you could find the greatest names in astronomy at work.

For 2 years, Kowal conducted brightness measurements on stars for Allan Sandage, a prominent cosmologist. It was fantastic. He spent night after night looking through the eyepieces of the 1.5-meter telescope on Mount Wilson or the 50-centimeter telescope on Palomar Mountain – splendid antique instruments with shiny tubes and brass cogwheels. After that he searched for supernovas – exploding stars in distant galaxies – for the opinionated Swiss astronomer Fritz Zwicky, one of the other eminent staff members at the Mount Wilson and Palomar Observatories.

Kowal liked Zwicky. He was a wonderful, brilliant man, but not what you might call diplomatic. If he did not agree with something, he made it clear in no uncertain terms. He made enemies a little too easily. But there was not a great deal to argue about in the search for supernovas or taking stock of remote galaxies and clusters, and Kowal continued doing the work, even after Zwicky's death in 1974. Between 1963 and 1984 he discovered 81 supernovas, one of which was bright enough to see with binoculars.

The dome of the Schmidt Telescope on Palomar Mountain, with which Chiron was discovered (Palomar Observatory/California Institute of Technology)

It was certainly painstaking work. Thirteen weeks a year, in the dark period around the New Moon, Kowal pretty much shut himself away in the dome of the 1.2-meter Schmidt Telescope on Palomar Mountain – which was in fact a giant telephoto lens with a large field of view, excellent for sweeping surveys of the night sky. He would place a photographic glass plate in the holder, start the exposure, spend an hour and a quarter looking through the guidescope to make sure the telescope continued to track the rotation of the sky, stop the exposure, store the glass plate in a lightproof container, and put in another plate for the next part of the sky. And then he would take the plates to Pasadena to examine them with the blink comparator, just like Clyde Tombaugh did in Flagstaff in 1930. He would compare the negative images with those of the previous month to see if, somewhere among the millions of black dots, there was a new star. And he did that for 20 years.

No wonder that, at the end of the 1960s, he started to become more interested in the solar system. On the many hundreds of Schmidt plates, countless asteroids had left short stripes on the photos during exposure. Normally Kowal would have ignored them, but if the trail of light was a little longer than normal, it showed that the asteroid in question was closer to the Earth, and these

Charles Kowal at the blink comparator used to discover Chiron (California Institute of Technology, courtesy George Beekman)

Earth-grazers were of course of great interest. At the request of comet and asteroid expert Brian Marsden from Harvard University, Kowal also searched for lost asteroids, whose position in the night sky could only be guessed at. In this way, between 1970 and 1980, he discovered just under 20 asteroids and six new comets.

Horace Babcock, director of Mount Wilson and Palomar Observatories, initially did not see much point in all that work on the solar system but, thanks to a letter from Marsden, he gave Kowal more or less a free hand. Once Charles had got into his stride, he also used the Schmidt Telescope to search for unknown satellites around the giant planets. That led him to discover a 13th moon around Jupiter, Leda, in 1974. A year later he identified another small satellite around the giant planet, which was not seen again until 2000 and is now called Themisto. With the most experience, the best wide-field telescope, and the most sensitive photographic emulsions, Charles Kowal was simply the right man in the right place at the right time.

Rhadamanthus

Before long, Kowal hit on the idea of searching not only for supernovas, Earth-grazing asteroids, and planetary satellites, but also for an unknown planet. After all, you never know. Beyond the orbit of Pluto, far from the Sun and therefore extremely faint, but perhaps within the range of the Schmidt Telescope. Planet X – that would certainly be something. Sure, Clyde Tombaugh had already scoured the entire night sky in the 1930s and 1940s from Flagstaff, but technology had advanced and the Schmidt photos could now reveal much fainter objects. Kowal even secretly thought of a name for the new planet he might discover: Rhadamanthus. The wise and immortal son of Zeus and Europa, who sat in judgment of the dead in the underworld, seemed a very appropriate choice.

He started on his new project in the summer of 1977. Over a period of about 10 years he should be able, alongside all his other work, to observe the entire zodiac, the broad band in the night sky within which the planets move. It would mean looking again in minute detail at many hundreds of Schmidt plates in the blink comparator. After all, new planets don't just fall into your lap.

Or do they? Early in November, just before Kowal's 37th birthday, he was examining photos taken on October 18 and 19, 1977 and discovered a faint point of light in the southwestern part of the constellation of Aries that had moved very slightly. It was clearly not an asteroid – it had not moved enough for that. During the exposure of the Schmidt plates the object had hardly moved at all. It had to be something in the outer regions of the solar system. Had he finally found Planet X?

'Don't run ahead of yourself,' warned Brian Marsden, who was in charge of the Central Bureau for Astronomical Telegrams at Harvard Observatory, the 'clearing house' of the International Astronomical Union (IAU), where new astronomical discoveries were registered. You cannot determine an accurate orbit on the basis of two or three observations and in theory it could also have been a new comet. The IAU announced the discovery on November 4 in circular 3129 to give other astronomers the opportunity to provide independent confirmation. And that came within a couple of weeks; the 'new' planet was found on old photographs from 1969 and 1941, and later even on a photo dating from 1895.

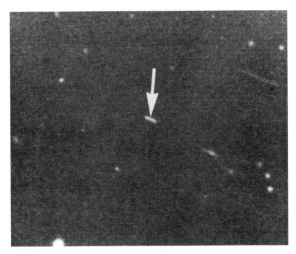

Discovery image of Chiron (California Institute of Technology, courtesy Charles Kowal)

Armed with all of these positional measurements, it was easy for Marsden to determine the object's path through the solar system. 'Object Kowal,' known officially as 1977 UB, proved to have a very extraordinary, elongated orbit, extending from just inside Saturn's orbit almost to that of Uranus. And the brightness measurements suggested that it was no larger than 140 kilometers across. What on Earth was this strange object? In Kowal's own words, it was 'too small to be a planet, too distant to be an asteroid, and it did not look like a comet.' 1977 UB defied astronomers' tendency to think in boxes, took no notice of the accepted labels, and seemed not to fit into any of the customary categories.

Centaurs

'Object Kowal' was a loner with little respect for conventions – and the search for the new planet had only been going on for a few months! It was logical to assume that there would be more of these strange 'mini-planets' drifting around in the outer regions of the solar system. Mother Nature makes bizarre things, but they are never one-offs. Kowal may not have found a full-fledged Planet X but 1977 UB, which seemed to be an unauthorized intruder in the territory of the solar system, was undoubtedly only the first of many of its kind.

But how should this new category of celestial body be classified? What kind of names should be given to these unauthorized mini-planets? Kowal did not have to think about it for long. During his lonely nights on Palomar Mountain he had read *The Centaur* by John Updike. It was a recognizable story of a difficult father–son relationship on the East Coast of the USA, based on the myth about the centaur Chiron, who gave up his immortality for the life of Prometheus, the bringer of fire.

Chiron – a wise and gentle healer – was the son of Cronos (known to the Romans as Saturn) and the grandson of Uranus. So it was an appropriate name for an object that spent most of its time moving between the orbits of Saturn and Uranus. The mythological centaur Chiron was neither horse nor human; the celestial body Chiron was neither asteroid nor comet. Or perhaps it was both. In any case, bodies similar to Chiron, which would undoubtedly be discovered, could be called centaurs. And there was certainly no lack of centaurs' names in Greek mythology.

Kowal started using the new name a couple of weeks after the discovery, before it had been officially approved by IAU. Journalists had difficulties with cryptic designations like '1977 UB'; 'Chiron' sounded a lot more acceptable. There was of course plenty of attention from the media, but it was as crystal clear that, even with the best will in the world, Chiron could not be called a planet: it was considerably smaller than the larger asteroids and its orbit could not be stable. Calculations showed that in 1664 BC, Chiron had passed within only 16 million kilometers of Saturn – just outside the orbit of its outer moon Phoebe. Similar encounters in the future are inevitable, resulting in radical orbital changes.

That could make life interesting for our distant descendants. If Chiron is thrown into a smaller orbit which carries it through the inner solar system, it will change into a super-comet. This strange object consists largely of ice and frozen gas, which will slowly but surely evaporate in the warmth of the Sun. In fact that is happening already. In 1988 Chiron suddenly became twice as bright – completely unheard of for an asteroid but not so peculiar for a comet. A year later, it was surrounded by a tenuous cloud of gas and dust comparable to a comet's 'coma.' Chiron is consequently one of the few celestial objects to have a dual classification: it is known as asteroid 2060, but also as comet 95P (the P stands for 'periodic').

Forbidden Zone

Remarkably enough, in 1989, when Chiron's resemblance to a comet was discovered, it was still the only known centaur in the solar system. After his amazing find in the fall of 1977 Kowal continued his unremitting search, but new discoveries eluded him. When he left the Mount Wilson and Palomar Observatories in 1985, after working there for 25 years, he had surveyed three-quarters of the zodiac, discovered a handful of new Earth-grazing asteroids and unknown comets, but had found no new celestial bodies in 'forbidden,' unstable orbits. Like Clyde Tombaugh, Charles Kowal found his pot of gold right at the start of his search and the time-consuming survey that followed produced little of significance.

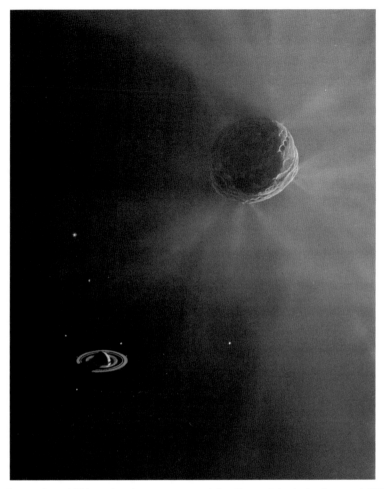

Artist impression of Chiron showing cometary activity. Saturn is in the background (William K. Hartmann)

A second centaur was not found until January 9, 1992, by David Rabinowitz using the Spacewatch Telescope in Arizona. Pholus, named after another wise and exceptional centaur from Greek mythology, had an even stranger orbit than Chiron. Sometimes it is closer to the Sun than Saturn, at others further away than Neptune. Its orbit is very elongated and at a very steep angle to the plane of the planetary orbits. Furthermore, Pholus is significantly larger than Chiron and extremely red in color. In fact, it is one of the most puzzling objects in the solar system.

Several dozen centaurs have now been found and there is no doubt that they originated in the Kuiper Belt – the wide band of small, icy bodies beyond the orbit of Neptune. Every one of them has an unstable orbit. In the distant future, today's centaurs will be expelled from the solar system as a consequence of the gravitational disturbances of the giant planets, or will be drawn into the inner regions, where they will grace the sky as impressive comets. But similar gravitational disturbances also draw new ice dwarfs inwards from the Kuiper Belt, meaning that the 'forbidden zone' between Saturn and Neptune will never be completely 'empty.'

And Kowal? He hoped that he would find Planet X and go down in history as the second American to discover a planet. That hope was never realized but, although Chiron is much too small to be considered a planet, the extraordinary centaur – half asteroid, half comet – is just as interesting to scientists interested in unraveling the evolution of the solar system. Reason enough to be proud – and Kowal never felt that pride more strongly than when he was awarded the James Craig Watson medal by the National Academy of Sciences in 1979.

The amateur astronomer from Buffalo, the lone observer, and supernova hunter, would never lose his fascination for the small objects in the solar system. As operational astronomer at the Space Telescope Science Institute in Baltimore – back on the East Coast – Kowal was one of the first to see the spectacular Hubble photos of the impact of comet Shoemaker-Levy 9 on the planet Jupiter in July 1994. And on February 12, 2001, as a programmer in the Applied Physics Laboratory of John Hopkins University, also in Baltimore, he witnessed the landing of the space probe NEAR-Shoemaker on the asteroid Eros.

In May 2006, Charles Kowal retired. He had spent a quarter of a century in California exposing, meticulously examining, and measuring photographic plates. For another 20 years he had been active in the much more technical world of space exploration, where the culture is much different and more commercial, and where he worked alongside wonderful, dedicated people. But now he was 65, he again felt the need to be alone with nature. For several months he traveled across the United States, visiting just about every national park in the country. Until, in Cinebar, Washington, in the foothills of the Cascade Mountains, he found his dream house. And for a dream price. Peace and quiet. And a lot of space.

Kowal tries to keep up with developments in astronomy, but he does not have a telescope, even though the night sky above Mayfield Lake can be stunningly beautiful. He has not seen Chiron, which is now in the constellation of Aquarius, for some time and it takes him some effort to recall all the details of the discovery. He is much more concerned about building a new robot, a fully controllable companion.

He is reluctant to answer questions about his youth and his failed marriage. That is too personal and no one else's business. Forbidden access for unauthorized intruders.

Chapter 10
Uranus: Problem Solved?

There's little love lost between Myles Standish and Tom Van Flandern. Back in the early 1960s, when they both studied at Yale University in New Haven, Connecticut, they got on well and respected each other. But Planet X has driven a wedge between them and ruined their old friendship. They no longer talk, avoid each other at conferences, and run each other down in interviews with journalists. And all for something that may not even exist.

Or perhaps it does. What about all those orbital disturbances in the outer regions of the solar system? Didn't they discover at the end of the nineteenth century that Uranus still displayed small orbital deviations that could not be explained by Neptune's gravitational pull? And wasn't that the main reason that Percival Lowell and William Pickering initiated a search for an unknown planet? A search that led to the discovery of Pluto in February 1930?

All of this is true. But Pluto cannot be Lowell's Planet X – the cold, remote object is much too small. That it happened to show up in the same part of the night sky where the mysterious planet was suspected to be must be a pure coincidence. Certainly after the discovery of the Plutonian moon Charon, which allowed Pluto's mass to be determined, it was clear that it could not possibly have any significant impact on the orbit of Uranus.

What, then, is causing the orbital deviations? They may be much smaller than the disturbances that led to the discovery of Neptune in the mid-nineteenth century, but there must still be a reason for them. Furthermore, there also appeared to be small deviations in Neptune's orbit, especially if you tried to match it to observations by Galileo Galilei in 1612 and French astronomer Michel Lalande in 1795. Without being aware of it, both of them had observed and recorded Neptune. As a result there were now positional measurements available for the planet dating back some 350 years.

So is there a Planet X floating around out there somewhere? Not if we are to believe Myles Standish, who now works at NASA's Jet Propulsion Laboratory in Pasadena. In May 1993 he published an article in *The Astronomical Journal* with the provocative title *Planet X: No dynamical evidence in the optical observations*. He concluded that there was no reason at all to assume that the solar system has a tenth planet. End of story, case closed. Except for a handful of astronomers, including Tom Van Flandern. In their view, it is easy for Standish

G. Schilling, *The Hunt for Planet X*, DOI 10.1007/978-0-387-77805-1_10,
© Springer Science+Business Media, LLC 2009

to say that, but in doing so he simply ignores the optical observations of many
seventeenth-, eighteenth-, and nineteenth-century astronomers.

Speculation

Tom Van Flandern (Courtesy Tom Van Flandern)

In 1963, after completing his postgraduate research at Yale, Van Flandern
started working at the Nautical Almanac Office (NAO) of the United States
Naval Observatory in Washington, D.C. For many years, together with its
British counterpart, the NAO has published the *Astronomical Almanac*, a thick
yearbook full of tables about celestial phenomena and planetary positions. To
be able to calculate these in advance, you need to know the orbits of the planets
very precisely. It was therefore very important for the astronomers at the NAO
to clear up any confusion about the motion of bodies within the solar system.

Bob Harrington (USNO, courtesy Jim Christy)

At the observatory on Observatory Hill, Van Flandern worked regularly with Bob Harrington, who also specialized in celestial mechanics. That often led to interesting projects and surprising publications, since anyone who keeps a close track on the movements of celestial bodies, soon encounters unexpected phenomena. And especially if you are not averse to speculative ideas which other astronomers might not wish to be associated with.

Tom and Bob, however, had little qualms in that respect. Once, they had studied supposed orbital deviations of Mercury and Venus and come up with the rather far-fetched theory that the small planet closest to the Sun had once been a moon of Venus, or rather that Venus and Mercury had once been a kind of binary planet. Mercury's escape could then explain the unusual rotation of Venus, which turns very slowly on its axis, and in the wrong direction.

Not that Harrington and Van Flandern lacked any critical capacity. Harrington, an ex-student of the Dutch–American astronomer Peter van de Kamp, had no time at all for the theory of his former mentor that Barnard's Star is accompanied by two Jupiter-like planets. Van de Kamp had reached this conclusion after many years of studying the nearby star, which seemed to show minute wobbles in the night sky. But when Harrington made more than 400 photographs of the star with the 1.5-meter telescope at the Naval Observatory in Flagstaff, he saw nothing to suggest any wobble. He concluded that Van de Kamp had been the victim of small systematic errors in his observation program – a conclusion that Van de Kamp contested until his death in 1995.

The question of Lowell's Planet X became topical again in 1978, when Jim Christy discovered Charon. Harrington was the first to calculate the mass of Pluto on the basis of the measured distance of Charon from Pluto and its orbital period. What everyone had suspected for a long time was finally established beyond any doubt: Pluto is by no means massive enough to cause deviations in the orbits of Uranus or Neptune. So what does cause them?

Humphrey

Around 1980 Harrington and Van Flandern both turned their attention to the problem of the tenth planet. Harrington approached it very pragmatically. To account for the observed orbital deviations of Uranus and Neptune you would need a planet at least five times as massive as the Earth. But a planet of that size should have been found long ago, for example, during Clyde Tombaugh's exhaustive search. Unless of course it is in a part of the solar system that has not yet been observed in great detail. For example, close to the south celestial pole.

Bob Harrington therefore had to assume that Planet X moved in a highly inclined orbit, almost perpendicular to those of the other planets. How a planet could end up in such a strange orbit would be something to worry about later. For the moment, it seemed interesting to find out whether the orbital deviations of the outer planets could be explained by a relatively massive Planet X which was currently located far to the south of the central plane of the solar system. Harrington, by the way, called his hypothetical planet 'Humphrey.'

At that time, Tom Van Flandern was 80% certain that Planet X was still hiding out there somewhere. He was aware that the positional measurements of Uranus and Neptune were much less problematic than they had been in the eighteenth and nineteenth centuries. But could that perhaps be because Planet X had a very eccentric orbit? Perhaps it was now much further away from the Sun than it had been a couple of centuries ago, with the result that the orbital deviations were less apparent.

Van Flandern tackled the problem slightly differently to Harrington. He collected as many positional measurements as possible and investigated whether they could all be explained by only one source of gravitational disturbance, be it a compact planet or a diffuse ring of dust and rubble outside the orbit of Neptune. But, to his amazement, this proved impossible: whatever celestial body or structure you pulled out of the top hat, you never managed to explain all the deviations. Apparently there was not one Planet X, but various unknown objects. Perhaps Harrington's Humphrey was one of them.

A targeted search for Humphrey, however, produced nothing and in the 1980s Planet X seemed more out of reach than ever, despite the fact that so many astronomers had addressed themselves to the problem. The Argentine astronomer Adrián Brunini even suggested in 1992 that the historical orbital deviations of Uranus were not caused by a tenth planet, but by a very recent collision with an object about 1,000 kilometers in diameter. That would explain why, at the end of the twentieth century, no more deviations were observed and the search for the tenth planet continued to be unsuccessful. These were desperate times for Planet X hunters, and they were resorting to desperate measures.

Exit Planet X

And these measures were a thorn in the side for Myles Standish. The Planet X hunters were always brandishing old optical observations, some of them centuries old when measurement errors of a few arc-seconds were no exception. We do not even know if Galileo tried to draw his rough sketches to scale, while positional measurements from the early twentieth century are equally suspect, since extremely accurate star catalogues were not yet available at that time.

These days, we can do a lot better. In the 1970s and 1980s, the Jet Propulsion Laboratory, NASA's control center for unmanned planetary research, determined the positions and orbits of the planets with much greater precision with the aid of modern techniques, such as radar echoes, radio-interferometry, and the measurements of planetary probes. And if you only looked at the modern measurements, there was no longer any evidence at all of inexplicable orbital deviations in the outer regions of the solar system. For Standish there was only

Myles Standish (Courtesy Myles Standish)

one obvious conclusion: the fact that there appeared to be deviations nearly a century ago was simply because it was more difficult to make accurate measurements.

Standish felt that, in a certain sense, the American space probe Voyager 2 had hammered the last nail into Planet X's coffin. The probe, launched in the summer of 1977, had examined Jupiter, Saturn, and Uranus close-up and, on August 25, 1989, it had a close encounter with the distant planet Neptune. The radio signals transmitted by Voyager enabled its position to be plotted with great precision, so the flight controllers in Pasadena knew exactly to what extent its course was affected by the planet's gravity. That enabled Neptune's mass to be calculated accurately. And the planet proved to be almost 1% less massive than had always been assumed.

It was time to revisit all the analyses of past decades. With more accurate planetary masses, better corrections for possible observational errors, more precise star catalogues, new orbital calculations, and, above all, without a prejudiced perspective. It needed to be an almost military operation – and Myles Standish was the right man for that. He was almost certainly a descendant of his seventeenth-century namesake, the army captain on the *Mayflower*, which carried the Pilgrim Fathers to America at the end of 1620, as the first European colonists. But where Standish senior helped conquer a New World on the other side of the ocean, Standish junior made short work of ideas about an unknown world beyond the orbit of Neptune. The latest data offered no justification at all for such an assumption.

A Pinch of Salt

All this of course went completely against the grain for Tom Van Flandern (Bob Harrington would probably have had the same reaction, but he died in early 1993 and never saw Standish's article in print). In Van Flandern's eyes, the optical observations of the past might not have been as good and as accurate as all of these modern techniques but you couldn't just push them aside that easily. Moreover, he and Harrington had once calculated that Neptune's strange system of moons could only be explained by assuming that it had once had a close encounter with a relatively massive planet. Who knows, it may have been the long-sought-after Planet X. No, that article by Standish deserved a little more critical comment.

Shouldn't Van Flandern himself provide that comment? But in the early 1990s, he was no longer taken seriously by his colleagues, and he had accepted his fate. In 1983 he had left the Naval Observatory to devote all his time to his revolutionary metatheory of the solar system, in which the planets originated as twins, exploded, and strewed asteroids and comets through space. In this theory, there was little room for conventional ideas about gravity. In 1993, 2 years after he set up his own organization Meta Research, Van Flandern's book *Dark Matter, Missing Planets and New Comets: Paradoxes Resolved, Origins Illuminated* was published. By that time he had largely used up all his goodwill in the world of planetary research and celestial mechanics.

According to Myles Standish, you have to take Van Flandern with a large pinch of salt. Definitely a capable man, but he has gone off in a very strange direction. An email message from Van Flandern to Standish's colleague Ron Hatch at the end of 1997, commenting on Standish' research into Planet X, is full of errors. And there is no sense at all in trying to put them all right, because Van Flandern will never accept that he has made mistakes. For Standish, this is typical of all those who believe in Planet X: instead of looking objectively at obvious solutions, they hang onto their idea of a tenth planet at all costs, just as UFO-believers see an extraterrestrial spaceship in every light in the sky. Standish even goes as far as to say that Brunini's 1992 article, which suggests that Uranus once collided with another celestial body, is a perfect example of scientific fraud, claiming that Brunini simply ignored any observations that did not fit in with his claims.

No, the two former fellow students will never make up their differences. In his turn, Van Flandern accuses Standish of being short-sighted, conservative, and obstinate. The evidence for his exploding planet theory is piling up, but no one is interested in it, at least not within conventional scientific circles. Since scientific journals refuse to publish his articles, Van Flandern publishes them on his own website. But he no longer believes that planetary experts will one day come to their senses and discard their own theories of the origin of the planets. They have become too set in their ways, blinkered in their ivory towers. What was that old saying again? Science advances funeral by funeral.

And what about Planet X? Is it still a viable concept? Van Flandern keeps all his options open. Perhaps the orbital deviations at the start of the last century were caused by large ice dwarfs in the Kuiper Belt, the band of celestial bodies beyond the orbit of Neptune. Perhaps they were caused by a number of different 'planets.' Who knows, perhaps there is a mysterious object out there, moving in a very elongated orbit. One thing is certain: we are nowhere near to having charted all the mass in the outer regions of the solar system, so there are still a lot of surprises in store. And if Planet X is ever found, there certainly won't be a look of surprise on Tom Van Flandern's face.

Chapter 11
Mysterious Forces

Pioneer 10 should not have changed course at all. After flying past the giant planet Jupiter, it left the solar system at a speed of 37 kilometers a second, heading for the star Aldebaran. It flew through empty interplanetary space, where there is no friction and its trajectory was determined by the gravitational pull of the Sun and the larger planets. The Pioneer Project's flight controllers could predict the position of the small space vehicle many years ahead. But Pioneer 10 did not keep to its designated route. In December 1992, it changed course very slightly, as if something had given it the lightest of taps.

It was years before the minute change was discovered, but eventually someone had to notice it. The probe's speed and position can be calculated with great precision from the radio signals it transmits to Earth. And when Giacomo Giampieri of Queen Mary and Westfield College in London took another look at all the old measurements, the effect was clearly visible. For more than 3 weeks a mysterious force had been affecting the small Pioneer, which was by then far beyond the orbit of Neptune. The force had not only knocked the probe slightly off course, it had also caused minimal changes in its velocity, first speeding it up a little and then slowing it down again. In the fall of 1998 Giampieri announced his discovery to the world. Pioneer 10 had apparently had a close encounter with a small, unknown planet.

For the second time in history a celestial body had been found in the outer regions of the solar system on the basis of orbital disturbances of another object. But this time, the disturbing body wasn't a large planet like Neptune and Giampieri was sensible enough not to mention Planet X. This could not possibly be the unknown planet which Lowell and Pickering had sought in vain. It could be no more than a hundred kilometers in diameter, otherwise its effect on Pioneer's trajectory would have been more noticeable, and it was not even visible from the Earth. Pioneer 10 had had an encounter with a small ice dwarf in the Kuiper Belt – the broad band of mini-Plutos discovered in the early 1990s.

For John Anderson, an expert on celestial mechanics at the Jet Propulsion Laboratory, it was a kind of deja vu, but in miniature. In the late 1970s, a couple of years after the launch of Pioneers 10 and 11, Anderson had realized that space probes could also serve as planet detectors. The gravitational pull of an unknown planet would cause a small but observable Doppler shift in the radio

G. Schilling, *The Hunt for Planet X*, DOI 10.1007/978-0-387-77805-1_11,
© Springer Science+Business Media, LLC 2009

signals transmitted by the Pioneers. If a large planet was still hiding somewhere beyond the orbit of Neptune, the Pioneers might be able to find it.

At the time, Anderson scarcely harbored any doubts at all that such a planet existed. In the early 1980s, the puzzle of the deviations in Uranus' orbit had still not been solved, while it was crystal clear that they could not be caused by Pluto. Everything seemed to point to a Planet X and Anderson thought that the ideas and searches of Bob Harrington and Tom Van Flandern deserved much more attention and appreciation. Who knows, perhaps the Pioneers would provide the decisive evidence.

Explorers

One thing is certain: the two space probes certainly did credit to their name. Like genuine pioneers, they explored unknown territory, paving the way for later probes like Voyagers 1 and 2. After the advent of space travel, at the end of the 1950s, it was not long before small unmanned spaceships were sent out to explore nearby celestial bodies at close quarters, first the Moon, but later also Venus and Mars, the Earth's two neighboring planets. And that was no easy task. Not only did these probes need to build up sufficient speed to escape the

Artist impression of NASA's Pioneer 10 spacecraft heading into outer space (NASA/Don Davis)

Earth's gravity, but also their trajectories had to be calculated so precisely that, months later, they would pass close to another planet. A planet that would also be racing along its orbit around the Sun at several tens of kilometers a second.

A journey to the giant planets in the outer regions of the solar system required even greater speed, otherwise the probe would fall prey to the Sun's gravitational pull before it had arrived safe and sound at its destination. That was why Pioneers 10 and 11 were so small and light and used powerful booster rockets when they were launched in March 1972 and April 1973, respectively. So the speed was no problem. But no one knew for certain if Pioneer 10 would ever get to Jupiter in one piece. To reach the outer regions of the solar system, it would have to pass through the asteroid belt, and just how much dust and rubble was flying around in there was anyone's guess. Pioneer could just as easily be sandblasted, dented, and riddled with holes by a cosmic hailstorm.

Jupiter, as imaged by Pioneer 10 (NASA/JPL)

In practice, everything went according to plan. In December 1973, Pioneer 10 flew past the giant planet Jupiter. A year later, Pioneer 11 equaled that feat and then flew on to the ringed planet Saturn, arriving in September 1979. The Pioneers' two larger brothers, Voyager 1 and Voyager 2, had been launched 2 years previously to study the giant planets with better instruments and more sensitive cameras. Voyager 2 even visited four giant planets: after flying past Jupiter and Saturn (in July 1979 and August 1981, respectively) it was the first probe to reach Uranus, in January 1986, and then, in August 1989, distant Neptune.

These flybys of the giant planets provided an abundance of new data, not only about the planets themselves but also about their extended systems of moons and rings. Sulfur volcanoes on Jupiter's moon Io, lightning discharges in

Saturn's atmosphere, 'shepherding moons' which control the dark rings of Uranus, nitrogen geysers on the Neptunian moon Triton – when it comes to new discoveries, the Voyager flights mark the Golden Age of planetary research.

Precision Measurements

Wouldn't it be wonderful if these mechanical explorers were also to find evidence of an unknown planet beyond the orbit of Neptune? They have certainly had every opportunity to do so: both the Pioneer and Voyager probes were traveling at such high speeds that they flew right out of the solar system – in different directions – on their way to the stars.

During that long journey beyond Neptune, the motion of a space probe is determined only by the gravitational pull of the Sun and the planets. Any encounter with an unknown Planet X would be detectable through a minimal change of speed and the slightest change of course. Precision measurements of the radio signals transmitted by a probe should reveal any changes of this nature. And especially in the case of the Pioneers, which turn on their axes a few times a minute, making them very stable as they fly through space. Extremely small increases in speed can therefore be determined much more accurately than with the Voyagers, which use thruster rockets and gyroscopes to maintain stability.

John Anderson (Courtesy John Anderson)

John Anderson knew that it would not be easy. And that he would need a large dose of luck. If Planet X were in the wrong part of the solar system, there would be no noticeable effects on the Pioneers' trajectories. So perhaps it was not so strange that they found nothing. But the orbital deviations of Uranus remained unexplained. And surely the articles by Bob Harrington and Tom Van Flandern were not just complete nonsense? In 1987, Anderson claimed that there probably was a Planet X, but that it moved in an elongated and steep orbit around the Sun. That would explain why Uranus displayed deviations at the end of the nineteenth century but has not done since. And if Planet X was now a great distance away above or below the Sun, you would not expect it to have any measurable effects on the trajectories of the Pioneers. According to Anderson Planet X would be four or five times more massive than the Earth and have an orbital period of around 750 years. It would not be observable again until around the year 2600.

Anderson's claims that Planet X still exists, despite there being no gravitational effect on the Pioneer probes was going a little too far for many other planetary scientists, especially after Myles Standish claimed in 1993 that the supposed orbital deviations of Uranus were based on observational errors. But Anderson's study of the Pioneer data did reveal another effect, which was, if anything, even more puzzling and for which there remains no conclusive explanation to this day: the two space probes had both lost speed, as if they had fallen under the spell of a mysterious force.

The Pioneer Anomaly

Experts in celestial mechanics are used to working with forces. Newton's laws of gravity know no secrets for them and, if they need to work at an extremely high level of accuracy, they can always apply Einstein's theory of relativity. In their research into the trajectories of the two Pioneer probes, Anderson and his colleagues used a force model – a kind of computer program taking account of all the forces acting on the probes. The most important forces are, of course, the gravity of the Sun and the major planets, but the model also includes the gravity of the smaller planets and the larger asteroids, as well as the extremely minimal effect of sunlight on the two Pioneers. All of these forces change continuously, for the simple reason that the solar system is constantly in motion. But the model takes all that into account so that, in theory, you can accurately work out the positions, directions, and speed of the two probes at any specific moment in the future.

In the course of the 1970s, however, the two Pioneers no longer appeared to be keeping to their celestial mechanical schedule and, in the early 1980s, the deviations were so significant that Anderson's team could no longer ignore them: there was clearly another force acting on the probes, a force that had not been incorporated in the model. Without including this unknown force in the calculations, it was impossible to explain the measured speeds and positions.

It was easy to jump to the conclusion that this unknown force could be explained by Planet X. That was Anderson's immediate reaction, too, but it was soon clear that the idea just didn't hold water. The mysterious force remained constant over the years and affected both Pioneers, despite their flying out of the solar system in totally different directions. It was as though some-thing were putting the brakes on the two probes in their journey to the stars. The effect was almost negligible, ten billion times weaker than the gravity on the surface of the Earth, but because it was constant, the consequences slowly but surely increased in significance. In 2007 the two Pioneers would be 400,000 kilometers closer to the Sun than they would have been without the mysterious force.

The most obvious explanation for this anomaly was that it was in some way caused by the probes themselves. Perhaps there was a tiny leak in one of the fuel pipes. Or the temperature distribution was slightly different than expected and the Pioneers radiated a little more heat in a certain direction. In both cases there would be a minuscule reactive force. But, after extensive study, Anderson and his colleagues came to the conclusion that this could not be the cause of the problem. It would be very strange if both the probes were subject to exactly the same force. Furthermore, a similar anomaly has also been observed in the radio signals from two other space probes, Ulysses and Galileo.

Perhaps the Sun is more massive than astronomers' thought? But that is impossible, otherwise the planets would have higher orbital velocities. Or maybe there is a cloud of invisible dark matter in the inner solar system? That is not possible either, as it would also have observable effects on planets, comets, and asteroids. Could something be going wrong in the analysis of the radio signals? No, because different teams have all detected the same effect with very different computer programs. Are the observed Doppler shifts perhaps caused by some unknown phenomenon in empty space? That is a possibility, but no one in the astronomical world has any idea what that phenomenon might be. Some physicists have suggested that there might be something wrong with current theories on gravity.

Planets Aplenty

It may be a long time before we find a solution to the puzzle. The two Pioneers have since stopped functioning, and Anderson and his colleagues have to make do with the data they have already gathered. There have been suggestions for a new space probe, specially designed to take extremely precise measurements of the unexplained effect, but that project still exists only on paper. It is therefore very doubtful whether John Anderson will ever live to see 'his' mystery solved.

It is all certainly very fascinating. In the mid-nineteenth century and the early years of the twentieth century observed gravitational disturbances led to searches for new planets. But in the case of the Pioneers, the exact opposite happened: the hunt for Planet X led to the discovery of inexplicable trajectory disturbances. Although, of course, some of them have been explained. Like in December 1992, when Pioneer 10 – then 7.8 billion kilometers from the Sun – felt the slight gravitational pull of an unknown object, resulting in a change of speed of a billionth of a millimeter per second. It was not Planet X, but almost certainly a small ice dwarf in the Kuiper Belt, one of many tens of thousands that must be in orbit in the cold outer regions of the solar system.

What other subtle forces will the two probes encounter in the future? No one will ever know since there has been no radio contact with Pioneer 11 since the mid-1990s and Pioneer 10 has been silent since 2003. But Anderson sees no reason why there should not be large planets orbiting the Sun at distances of many billions of kilometers – in extremely slow orbits, through bitter cold and darkness. After all, no one knows how far the solar system extends and what mysterious objects may still be hiding in its outer regions.

You never know, perhaps Pioneer 10 will some day end up on a planet in a completely different solar system. During its journey among the stars, the tiny probe will suffer almost no wear and tear. It will certainly outlive not only its builders, but also humanity as a whole and perhaps even the Earth. Who knows where it will finally come to rest as it journeys through the Milky Way. After all, there are planets aplenty.

Chatper 12
The Hunt for the Death Star

Richard Muller sighed deeply once again and took yet another long look at the diagrams and formulae on the large blackboard. How does nature manage to bombard the Earth with asteroids every 26 million years? Together with his colleague Marc Davis, Muller had investigated just about all the possibilities. But to no avail. There was still no explanation for the periodic asteroid showers.

At that moment, he heard light footsteps in the long, high corridor of the University of California at Berkeley. A small man stuck his head around the door of Muller's room. He had long, wavy hair down to his shoulders and a massive beard. Piet Hut was in Berkeley for 2 days to attend the wedding of his colleague Simon White and of course he had to say Hi to the scientists he had met 6 months earlier, while staying at the university for a few months in the summer. As curious as ever, he asked Muller and Davis what they were doing.

'Have you ever thought it might be a periodic shower of *comets* rather than asteroids?' Hut asked when they told him about their problem. 'Look, perhaps it works like this. Give me the chalk a minute.' A rough sketch, a bunch of new formulae, a couple of quick calculations and, within half an hour, the Nemesis theory was born. Of course, they still needed to work out a few details, which they did later at Muller's home. The following morning they posted an article to *Nature* and Hut flew back to Princeton.

That was the end of December 1983. In April the following year, Piet Hut was back in Berkeley again staying with friends, and someone showed him the *San Francisco Chronicle*. For Hut, Nemesis had been a minor episode in a multifaceted career, but now – only days after the article was published in *Nature* – the story about the Death Star was on the front page of the newspaper and the telephone was red hot with calls from press and television journalists. A mysterious companion of the Sun which fired comets at the Earth every 26 million years, reaping death and destruction across the planet – that is news that fires the imagination.

G. Schilling, *The Hunt for Planet X*, DOI 10.1007/978-0-387-77805-1_12,
© Springer Science+Business Media, LLC 2009

Dinosaurs

The seeds of the Nemesis theory had already been sown in 1980. In that year, Luis Alvarez, the world famous nuclear physicist and Nobel Prize winner, and his son Walter published an article claiming that an asteroid had collided with the Earth 65 million years ago, ending the age of the dinosaurs. Throughout the world, rock strata dating back to that era contain unexpectedly high levels of iridium – an element that is very rare in the Earth's crust but relatively common in meteorites and asteroids. The impact of a piece of space rock about 10 kilometers in diameter had supposedly caused a radical change in the Earth's climate. For many years the Sun was obscured by clouds of dust, food chains were disrupted, and 90% of all biological species in the oceans died out. On land, the impact was the death knoll for the dinosaurs, who had ruled the Earth for more than 150 million years. The impact had one small but far-reaching consequence – the extinction of the giant reptiles allowed mammals to evolve more quickly. Without the asteroid impact, man would probably never have existed.

In Berkeley, Luis Alvarez had been Richard Muller's tutor. Muller started as a particle physicist but soon became more interested in Earth sciences. And although most geologists were exceptionally skeptical about the impact theory of Alvarez and his son, Muller was convinced almost immediately. It had always been assumed that climate changes and mass extinctions were caused by geological processes, like large-scale volcanic activity or changes in ocean currents. But

A giant impact may have ended the age of the dinosaurs, 65 million years ago (NASA/Don Davis)

Richard Muller (right) with his Berkeley tutor Luis Alvarez (Courtesy Richard Muller)

the thin layer of iridium deposited on the surface of the Earth 65 million years ago suggested conclusively that the biological catastrophe that occurred in the transition period between the Cretaceous and Tertiary eras had an extraterrestrial cause. Who knows: perhaps other mass extinctions were also caused by cosmic projectiles.

Muller was therefore also exceptionally interested in a sensational article by two paleontologists at the University of Chicago, published 3 years after the advent of the impact theory. David Raup and John Sepkoski had conducted extensive research into biodiversity in the geological past, especially in the oceans. They came to the disconcerting conclusion that mass extinctions appeared to happen at astoundingly regular intervals. According to Raup and Sepkoski, there had been about 10 in the past 250 million years, at intervals of 26 million years.

At first, Sepkoski and Raup could hardly believe their own results. There was no reason at all to assume that large-scale changes on Earth should occur with such strict regularity. But Muller made a direct link with the Alvarez impact theory. After all, there is no branch of science where regularity and periodicity play such an important role as astronomy. Suppose most mass extinctions had an extraterrestrial cause, was it possible that the Earth was bombarded with asteroids every 26 million years?

And that was why, on that December afternoon in Berkeley, in the hills to the east of San Francisco, Muller was trying to work out, together with Marc Davis, how nature succeeded in shooting off a salvo of celestial curtain fire at the Earth every 26 million years. But they had no luck – until the small, long-haired genius from Princeton dropped in and suggested that it was not an asteroid shower but periodic bombardment by comets.

Nemesis

Piet Hut studied physics and astronomy in Utrecht, in the Netherlands, and wrote his dissertation on the quantum effects of black holes. But he was equally interested in Eastern philosophy and spirituality. In everything, in fact. He found it so difficult to focus on one single subject for his PhD research that, at the end of the 1970s, he switched to the University of Amsterdam, because there he would be able to combine different subjects. Hut's thesis was in fact nine different articles stapled together and covering topics from neutron stars and cosmology to the celestial mechanics of binary stars. Amsterdam astronomer Ed van den Heuvel thought Piet was the most brilliant student the astronomical institute had ever encountered, and was very sorry

Piet Hut now uses dedicated supercomputers to tackle astrophysical problems (Courtesy Stichting De Koepel)

when Hut left in 1981 to go and work at the renowned Institute for Advanced Study in Princeton.

Understanding a problem at a glance, thinking outside the box, offering a solution, working out the details, and then moving on to the next one – that was typical of Piet Hut. Of course, it helped that he had just developed methods for calculating the dynamics of globular star clusters, in which the motion of each star is influenced by the gravity of hundreds of thousands of others. The same math proved to be applicable, with some small adjustments, to the dynamics of the Oort Cloud, an enormous cloud of trillions of comets at a great distance from the Sun. Imagine that the Sun is actually a binary star with a small, faint companion in an elongated orbit with an orbital period of 26 million years. Once per orbit, the companion would pass through the Oort Cloud, disturbing the orbits of hundreds of millions of comets. A large percentage of them would enter the inner solar system, causing more cosmic impacts on Earth than normal.

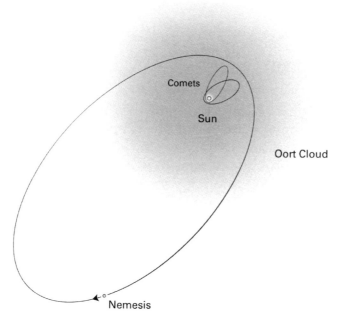

In its eccentric orbit, the Sun's hypothetical companion Nemesis would pass through the Oort Cloud and send comets into the inner solar system (Wil Tirion)

Muller, Hut, and Davis called the Sun's hypothetical companion Nemesis, after the Greek goddess of divine retribution. It is not as crazy as it sounds to suggest that the Sun is a binary star. Two-thirds of all stars are either binaries or part of a multiple system. But wouldn't a companion to our Sun have been discovered long ago? Not necessarily: if it is a faint red dwarf star, it would not be easy to observe. It may even have been seen and photographed countless times without anyone realizing that it is our Sun's nearest neighbor. And the Nemesis theory explains why most mass extinctions have happened reasonably gradually rather than suddenly: a comet shower can last hundreds of thousands of years, during which time an estimated 20–30 impacts occur. The theory even explains small variations in the 26 million year cycle: although Nemesis orbits the Sun, it is also affected by the gravitational pull of other nearby stars, which may slightly disturb its orbit. On that December afternoon, everything fell into place and the *Nature* article was written in the twinkling of an eye.

Of course, Davis, Hut, and Muller were not the only ones to offer an explanation for the periodic mass extinctions discovered by Raup and Sepkoski. On April 19, 1984, *Nature* published something of a theme issue on the question, with five articles and two comments on three different hypotheses. NASA scientists Michael Rampino and Richard Stothers described how the Oort Cloud could also be disturbed by the Sun's wave-like motion through the Milky Way: sometimes the Sun is a little above the central plane of the Milky Way and sometimes a little below it. Once every 33 million years it moves through the central disc, which contains large, massive clouds of dust and cool gas. The gravity of these molecular clouds would also cause disturbances in the Oort Cloud. Richard Schwartz and Philip James of the University of Missouri presented a similar hypothesis but Muller and Davis thought the Milky Way theory was full of holes. Molecular clouds may have a large mass, but they are very extended and tenuous, and therefore have little effect on the motions of comets in the Oort Cloud. In their view, the theory offered by Daniel Whitmire and Albert Jackson was much more plausible. Whitmire, a physicist at the University of Louisiana in Lafayette, and Jackson, from the Computer Sciences Corporation in Houston, presented a 'light' version of Nemesis. They suggested that the Death Star is not a red dwarf, but a much smaller, lighter, and fainter *brown* dwarf – a failed star not much larger than Jupiter. And it would have a very elongated orbit, passing through the inner parts of the Oort Cloud with each revolution.

Questions

Nemesis was hot but, as the year progressed, there were a lot of questions about the Death Star theory. Was the orbit of the Sun's hypothetical companion stable? At such an enormous distance – more than a light-year – a low-mass red dwarf could easily be wrenched away from the grip of the Sun's gravity. On October 18, 1984, *Nature* published a new series of articles on Nemesis (including one by Piet Hut) which addressed this question in detail. Nemesis' proposed orbit did indeed appear to be unstable in the long term. In an accompanying comment entitled 'Nemesis for Nemesis?,' Mark Bailey of the University of Manchester suggested that Hut's article should be seen as a 'near retraction' of the Nemesis theory.

This astounded Hut and enraged Muller. The fact that Nemesis' current orbit was unstable might imply that the Sun would lose its red dwarf companion in a billion years or so, but that did not mean that the Sun could not have originated as a binary star. Hut's calculations showed that such a widely separated binary would become increasingly unstable over time, but its estimated life expectancy said nothing of the past. Muller protested to Bailey, who claimed resolutely that he had not used the term 'retraction' himself and that it must have been added by the editors at *Nature*.

The upshot was, however, that no one believed in Nemesis anymore. They all remembered Bailey's critique, while few people took the time to read Hut's original article, which he himself considered as supporting the Nemesis theory. All too often Muller receives the startled response: 'Nemesis? Wasn't that idea refuted long ago?' To which he always answers: 'By whom?' As far as Muller is concerned, 20 years after its birth, Nemesis is still alive and kicking. Doubts may have arisen recently about the periodicity discovered by Raup and Sepkoski (statistical significance is a tricky subject among scientists), but mass extinctions over the past 250 million years have by no means occurred randomly. Furthermore, a comparable periodicity has been discovered in the ages of large impact craters on Earth. So there would certainly seem to be something in the whole idea.

So why has Nemesis still not yet been found? Muller puts this down to the fact that we simply have not put enough effort into looking for it. He himself made an attempt in the 1980s, as a sort of spin-off of a supernova search program being carried out by his student Saul Perlmutter (which would later lead to the discovery of the accelerated expansion of the universe). But the telescope he used was not up to the immensity of the task. And although the distances of a large number of faint stars have now been determined, it is still not 100% out of the question that somewhere in the night sky there might be a red dwarf orbiting the Sun at a great distance. 'Give me a million dollars and I'll find it,' Muller once said to a journalist on the American website space.com.

A Remote Planet X

A red dwarf is not a planet and Nemesis has therefore never been identified as Planet X. But the search for the Death Star has much in common with the hunt for the tenth planet. Couldn't the periodic comet showers just as easily be caused by a distant, unknown planet? Do you really need a red dwarf to explain them? In 1984, Daniel Whitmire claimed that the phenomenon could be caused by a brown dwarf – a warm ball of gas somewhere in between a dwarf star and a giant planet. And in 1999 he adjusted the mass of his 'Nemesis-light' even further downward, to three times the mass of Jupiter.

Whitmire had joined forces with two colleagues from Louisiana, John Matese and Patrick Whitman. Matese studied the orbits of more than 80 long-period comets. These are comets with orbital periods of more than 200 years, in many cases even tens of thousands of years. He discovered that their orientation in space was not completely random. A similar effect had also been observed by John Murray of the Open University in Milton Keynes, England. It seemed as though the comets had been forced into their current orbits by the disturbing effect of a massive planet about a thousand times further away from the Sun than Pluto. And, according to Matese, Whitman, and Whitmire there were probably many more.

That was the last time in the twentieth century that Planet X made the newspapers. Of course, everyone wanted to know where the distant planet was exactly, how it could be found, and what it was to be called. But both Murray and Matese were smart enough not to commit themselves in any way – after all, their argument was not that strong. As long as the planet had not actually been photographed, nothing definite could be said about it and it remained a completely hypothetical object.

In the meantime doubts have been raised about the analyses of the two teams. According to Jonathan Horner and Wyn Evans of Oxford University, the catalogues of cometary orbits on which Murray and Matese based their calculations did not give a representative picture of the situation in reality. They contain, for example, more comets that are visible from the northern hemisphere, because that is where most astronomers are located. Furthermore, a winter comet has more chance of being discovered than a summer one, because then the nights are longer. In their view, all these selective effects left little intact of the statistical analyses of the two teams.

Piet Hut in any case no longer devotes much attention to the problem. He agrees that Nemesis might exist and a distant Planet X is also not completely impossible, but there is no real reason to believe in them any more. Not that he regrets his Nemesis adventure in the 1980s: it brought him into contact with geologists and paleontologists, and there is nothing better than interdisciplinary research. At the Institute for Advanced Study he regularly sits around the table with theoretical physicists, biologists, computer scientists, and philosophers, and he has good memories of his meeting with the Dalai Lama in 1997. But when it comes to cosmic impacts, Hut stays closer to home. He is one of the founders of the B612 Foundation, named after the asteroid home of the Little Prince in the book by Antoine de Saint-Exupéry. The foundation conducts research into the possibility of deflecting an Earth-crossing asteroid and thereby averting a cosmic catastrophe. A much more urgent issue than whether Nemesis exists or not.

And that makes Richard Muller in Berkeley one of the few scientists who is still seriously occupied with the Death Star. As recently as 2002 he published the results of measurements made on microscopic glass pellets in samples of moon rock brought back to Earth in 1971 by the Apollo 14 astronauts. He claims that the results show how the number of impacts on the Moon suddenly increased some 400 million years ago, possibly because Nemesis found itself for the first time in an orbit that periodically took it through the Oort Cloud. It would be a shame to discard such a wonderful theory just because it had gone out of fashion.

No, that will have to wait for a while until the evidence against Nemesis is conclusive, for example, when future searches with large, automated telescopes continue to produce no clear results. The issue should be cleared up by around 2010. But the question of how the periodicity of the mass extinctions can be explained will remain unanswered. Maybe, around that time, Piet Hut should drop by again.

Chapter 13
The Secret Planet

Planet X? That was discovered long ago. But NASA is keeping it secret, undoubtedly in close collaboration with observatories throughout the world. The planet goes around the Sun every 3,600 years in an elongated orbit and is currently on a collision course with the Earth. The disastrous portents of that imminent encounter – probably in 2012 – are visible everywhere. And that Nibiru, as Planet X is called, is being kept under wraps by the secret world government is not so surprising since the apocalyptic celestial body is populated by the Anunnaki, the highly developed civilization that was responsible for creating humanity.

Planetary scientists are being driven to distraction by Nibiru. As are archaeologists. And it is not surprising – you devote so much time, energy, and creativity to fascinating scientific research and find yourself on the tracks of the most amazing and interesting things and all the public at large is concerned about is some crackpot theory about clay tablets, god-astronauts, and a planet that doesn't exist. And when you try to explain why it is all just a fabrication, you end up on the believers' blacklist as being in the employ of NASA or the CIA.

Zecharia Sitchin

G. Schilling, *The Hunt for Planet X*, DOI 10.1007/978-0-387-77805-1_13,
© Springer Science+Business Media, LLC 2009

Nibiru is the creation of the American Zecharia Sitchin, who wrote a bestseller called *The Twelfth Planet* in 1976. In Sitchin's theories, the planets bounce around cheerfully in space, with all kinds of catastrophic consequences. This is nothing new of course: the Russian Immanuel Velikovsky made similar claims in 1950 in his controversial book *Worlds in Collision*. Velikovsky believed that the Earth had a close encounter with Venus a few thousand years ago, and that Venus was a comet that had originated as part of Jupiter. Original to say the least.

Sitchin combines Velikovsky with Erich von Däniken. And why not? In 1968, the Swiss author wrote the bestselling *Chariots of the Gods?*, in which he claims that just about all religious myths in the history of mankind can be traced back to visits to the Earth by extraterrestrial beings. It was pseudoscience of the highest order, and it sold like hot cakes. A bit of Velikovsky, a bit of Von Däniken, a dash of biblical exegesis, and a pinch of New Age – that is the recipe for Sitchin's Nibiru pie.

A Ready Pen

There seems to be no end to these nonsense stories and conspiracy theories. The pyramids were built by extraterrestrial civilizations. Lines in the ground in Peru are runways for spaceships. Atlantis was a launchpad. Sodom and Gomorrah were destroyed by a nuclear explosion. There is an artificial human face on the surface of Mars. Crop circles are made by UFOs. And, of course, the Apollo landings were faked, the government secretly adds fluoride to our drinking water, and the World Trade Center was destroyed to justify the wars in Afghanistan and Iraq. It's enough to wear you out.

As soon as people turn their attention to lost civilizations, these kinds of theories come to the fore. They all follow the same pattern. You pluck an idea completely out of the air and link it suggestively to something about which little is known with any certainty. If you then select the available facts carefully and have a ready pen, speculative interpretations can soon sound very convincing.

Sitchin certainly has a ready pen. His *Earth Chronicles* now extend to six volumes. And he is certainly selective in his use of the facts, at least according to his many skeptical critics. His thousands of fans do not seem to care that he has no understanding of either Sumerian and Akkadian grammar or celestial mechanics. Nor does it matter that there is no sign at all of the mysterious planet Nibiru, despite all the predictions.

Sitchin, who was born in Russia in 1922 and grew up in the Middle East, studied economics in London and taught himself to read and interpret Sumerian clay tablets. His Nibiru theory is based on an image on a cylinder seal, showing 11 small spheres grouped around a sort of star. Sitchin quickly jumped to the conclusion that this must be a map of the solar system. Apparently the Sumerians knew that there were more planets than the seven (including the

The Sumerian cylinder seal interpreted by Zecharia Sitchin as a crude map of the solar system

Sun and Moon) known in the ancient world. But, wait a minute: if the Sun is the star in the middle that makes *12* in all. So the civilization on the banks of the Euphrates was clearly aware of the existence of an as yet undiscovered planet!

When Sitchin took a close look at the Sumerian creation myth *Enûma Elish*, everything fell into place. The gods in the epic were in fact celestial bodies. Marduk (or Nibiru), which moves through the solar system in an elongated orbit, long ago collided with the unfortunate planet Tiamat, creating the Earth, Moon, and asteroids. And for good measure, during another circuit it catapulted the Saturnian moon Gaga (alias Pluto) out of its orbit.

And that is not all. Nibiru is the home planet of the Anunnaki, the gods from the *Enûma Elish*. They had set their sights on the Earth's gold and landed on our planet half a million years ago. They needed a slave race, so they created humanity 'in their own image and likeness' using genetic manipulation. This is the background to the sons of God and giants described in Genesis 6. By the way, the tower of Babel – a tower that reached up to heaven – was of course a space ship. The fact that Genesis 11 clearly states that it was made of clay bricks and pitch does not seem to matter.

The flood described both in the Bible and the Sumerian Gilgamesh epic, was also the work of the inhabitants of Nibiru. They wanted to punish mankind and flooded the Earth by breaking off a piece of Antarctica. They escaped destruction themselves and returned to their home planet, which passes through the inner solar system every 3,600 years. Sometimes that is good news for mankind – for example, the visitors may provide instruction in agricultural techniques – while at others, it is not so favorable, such as with the destruction of the Minoan empire in 1650 BC.

Conspiracy

How do you respond to such nonsense? You could just shake your head and ignore it, but most astronomers are not prepared to go that far. Scientists are used to arguments and deductions, reason and logic. They cannot imagine that the rest of mankind considers these methods less important. So they write complete websites debunking the Nibiru myth, they point out the weak points in Sitchin's argument, and make it abundantly clear that he has no understanding whatsoever of astronomy. All a complete waste of time, of course, because none of this will have any effect on 'real' believers. On the contrary, they see it as a conspiracy to conceal the truth.

Moreover, every time an astronomer uses the term 'Planet X,' they see it as proof of the fact that there really is something going on and that it is only a matter of time before Nibiru is discovered. In 1983, for example, the *Washington Post* reported that the American–Dutch infrared satellite IRAS had discovered a tenth planet in the solar system. That was jumping the gun a little: IRAS scientists Gerry Neugebauer and James Houck had reported at a press conference that the satellite had detected mysterious sources of infrared radiation which could be anything, from distant galaxies to unknown planets. Eventually, the radiation indeed turned out to be from remote, dusty galaxies, but Sitchin's followers know for sure that it was a cover-up and that Neugebauer and Houck accidentally let slip that Nibiru had been discovered.

And in February 2000, when a distant celestial body was discovered and catalogued as 2000 CR_{105}, all hell broke loose. Although the ice-dwarf has an orbital period of just over 3,000 years and an elongated orbit, it could not possibly be Nibiru because it never comes closer to the Sun than Pluto. But Brett Gladman, then associated with the Nice Observatory in France, wrote in the journal *Icarus* that the peculiar orbit of 2000 CR_{105} could possibly be caused by the disturbing influence of an unknown Planet X. The planet would be about as massive as Mars and orbit the Sun at an average distance of 15 billion kilometers. A report in the American weekly *Science* on April 6, 1991 (written by the author of this book) is still cited regularly on Nibiru websites as 'scientific evidence' of the existence of the tenth planet.

And it gets even crazier. When the Griffith Observatory in Los Angeles, one of the largest public observatories in the United States, closed its gates to the public on January 6, 2002, Nibiru believers saw it as the evidence that there was something to hide. Perhaps the powers-that-be wanted to ensure that the observatory's visitors did not catch a glimpse of the approaching planet? It was an insane idea, but it fits in with the equally insane conspiracy theory. In reality, the observatory was undergoing an extensive renovation and expansion project. Its doors opened again in November 2006.

Zetas

This was all, by the way, after the people of Earth had been informed of the existence of Planet X by an extraterrestrial civilization. The Zetas, beings with large eyes who inhabit a planet in orbit around the star Zeta Reticuli, made themselves very clear: Planet X was on the way. It would pass very close to the Earth and cause a catastrophic pole shift. This was all made known during the 1990s. Not by radio signals from the cosmos, but channeled through medium and Zeta envoy Nancy Lieder, who was in contact with the altruistic aliens.

And Lieder had other important news from Zeta for her credulous disciples: comet Hale–Bopp, which was due to make a spectacular appearance in the night sky in the spring of 1997, did not actually exist! It was a diversionary maneuver dreamed up by NASA to distract attention from the rapidly approaching Planet X. Lieder predicted that Hale–Bopp would not show and that NASA would cook up some excuse about it breaking up unexpectedly.

Comet Hale–Bopp in the spring of 1997 (Robert Wielinga)

Lieder's predictions can no longer be found on her website zetatalk.com. This is not surprising, since Hale–Bopp was by far the most conspicuous comet of the past 25 years. Even city-dwellers standing in their illuminated living rooms could see the impressive 'star with a tail' shining in the night sky with the naked eye. The fact that it clearly existed, however, did little to allay the myth and superstition surrounding Hale–Bopp. The founders of the Heaven's Gate sect, for example, claimed that there was a UFO flying in the wake of the comet, which would pick up the select group of followers and carry them off to paradise. That belief led to the tragic collective suicide of 38 members of the sect.

After the Hale–Bopp fiasco, around the millennium, Nancy Lieder announced new revelations from the Zetas. Planet X would pass close to the Earth in the spring of 2003, causing its axis to tilt by 90 degrees, wiping out 90% of humanity. Lieder also predicted similar catastrophic events on May 5, 2000, when several planets in the solar system were more or less aligned. And, although Zecharia Sitchin has never referred to Lieder's predictions, his

Logo of the Heaven's Gate sect

followers had no scruples about linking the Zetas' channeled communications to 'their' Nibiru.

Portents of disaster and doomsday scenarios started to crop up everywhere, and dozens of websites told us exactly what apocalyptic events were in store for all of us. And, of course, the influence of the approaching planet was clear wherever you looked. The hole in the ozone layer, El Niño, the unpredictable behavior of the Sun, record-breaking climatic phenomena, an alleged increase in the number of earthquakes – everything was seen as evidence that the mother of all cosmic dramas was at hand. No one seemed surprised that the planet, which should have long been within the orbit of Saturn, was still not visible in the night sky.

Mayas

The End of Days has been predicted on many occasions and the prophets of doom have never been right yet. And in 2003 they were wrong again. Like any other, that year had its share of disasters and dramatic events. Seven astronauts in the space shuttle Columbia were killed on re-entry into the Earth's atmosphere, the SARS epidemic broke out, and the European Mars landing craft Beagle 2 came to grief on the Red Planet. The year 2003 was also the year in which the human genome was first fully mapped, Mars reached its closest point to the Earth in 50,000 years, and director Peter Jackson completed his movie trilogy *The Lord of the Rings*. But there was no sign at all of a planet approaching the Earth, a tilting axis, or a new Ice Age.

So what now? Do Nibiru or Planet X not exist after all? Had the Zetas got it wrong? Or could the myth still survive? Of course it could – when have the supporters of a pseudoscientific idea ever admitted that they have hold of the wrong end of the stick? For his part, Zecharia Sitchin has never specified a date on which Nibiru would once again visit the Earth. And Nancy Lieder has now been informed that the apocalypse will now probably not take place until 2012. That was a clever move, because it immediately attracted the attention of Maya cranks, who have known all along that something special is due to happen on December 21 of that year.

The Central American Maya culture had a complicated calendar, which used what is known as the 'long count' system. One cycle consists of 13 *baktun*. One *baktun* (144,000 days) is 20 *katun*, one *katun* is 20 *tun*, one *tun* is 18 *uinic*, and one *uinic* is 20 *kin* (20 days). The current cycle started on August 11 in the year 3114 BC. A simple sum therefore shows that it will come to an end on December 21, 2012. And, according to some Maya myths, that will coincide with the re-creation of the world.

So that means there is plenty to do for the debunkers – the archaeologists and astronomers who take a long and skeptical look at the tidal wave of Nibiru nonsense and explain with scientific precision what is wrong with this cosmic fairy-tale. They will have their work cut out in the next few years. And on December 22, 2012 there will be a new pseudoscientific cock-and-bull story doing the rounds and the whole circus will start all over again. Because no matter how many new celestial bodies are found in our solar system, there will always be a need for a mysterious Planet X.

Chapter 14
Vulcanoids and Earth-Grazers

Dan Durda really had to do something about his weight. So it was off to the gym, for some serious training. But it was all in a good cause, otherwise he could forget about flying an F-18. It wasn't that Durda was too heavy – on the contrary, he was 3 kilos too light. To qualify for the ejector seat in the fighter jet, you had to weigh at least 63 kilos. Durda – 5 feet 11 in his socks and slightly built – weighed in at only 60 kilos. It was high time to build up some muscle power.

Daniel David Durda has always got a kick from anything fast and furious. And maneuverable, of course. That's why he loves racing around in his bright red Chevrolet Camaro. And why he loves flying, not only Cessnas from the Davis-Monthan base in Tucson, Arizona, but also these days F/A-18Bs from Edwards Air Force Base in California. And if he feels the need for some peace and quiet, he goes hiking in the Rocky Mountains or cave diving in Florida, or he paints space art at his home in Boulder, Colorado.

Dan Durda, flying an F-18 at high altitude (Courtesy Dan Durda)

It's an added bonus that flying in the F-18s also has some scientific value. Shortly after sundown, Dan Durda takes off, climbs to 50,000 feet – where the

G. Schilling, *The Hunt for Planet X*, DOI 10.1007/978-0-387-77805-1_14,
© Springer Science+Business Media, LLC 2009

sky is deep indigo – and switches on his ultraviolet wide-angle camera to look for vulcanoids, hypothetical celestial bodies that circle the Sun within the orbit of Mercury. Each flight costs around 10,000 dollars. Not a lot of money for a handful of mini-planets – which haven't been actually been found yet.

Vulcan

Searching for intramercurial planets from inside a metal bird soaring around high above the Earth at 90% the speed of sound is something Urbain Jean-Joseph Le Verrier could never have imagined. In December 1859, the French astronomer – famous for successfully predicting the position of Neptune, which led to the planet's discovery – was himself searching for a planet within the orbit of Mercury. But he wasn't high above the Earth, he was in a carriage in the village of Orgères-en-Beauce, 100 kilometers to the south of Paris.

On December 22, Le Verrier had received a letter from Edmond Lescarbault, who was the village physician in Orgères and an avid amateur astronomer. He asked whether the director of the Paris Observatory could take note of his observations, made earlier that year, which appeared to show that there was a planet circling the Sun within the orbit of Mercury. Le Verrier wanted to know more. He had spent nearly 20 years trying to understand Mercury, the closest planet to the Sun. Mercury's orbit is quite eccentric, and the orbit's orientation rotates very slowly: the perihelion, the point at which the planet is nearest to the Sun, only shifts gradually. According to Le Verrier, 90% of the perihelion shift was caused by gravitational pull of other planets, but he could find no satisfactory explanation for the remaining 10%. Unless there was another planet moving within Mercury's orbit.

Once you have successfully predicted the existence of an unknown planet on the basis of observed orbital deviations, the same explanation will obviously spring to mind the next time it happens. Had the country doctor really seen the planet that could solve the Mercury problem? Le Verrier wasted no time and traveled straight to Orgères with a colleague to talk to Lescarbault in person.

It was a strange encounter, because the Parisian astronomer did not at first say who he was. Only after Lescarbault had told the mysterious visitor in detail what he had seen on March 26, 1859 did Le Verrier reveal his identity. It all sounded very convincing. On that spring day, the experienced amateur observer had looked through his telescope and seen a small round dot move across the face of the Sun. It couldn't have been a sunspot, because these cooler areas on the Sun's surface do not move that quickly. It looked very much like the transit of a planet. Just as when Mercury had revealed itself in 1848, it was a small black pellet, the silhouette of a planet, passing every now and then exactly between the Earth and the Sun.

Le Verrier examined Lescarbault's notes and calculated that the intramercurial planet must have an orbital period of just over 19 days, circling the Sun at an average distance of 21.4 million kilometers. There was only one problem: the dot that Lescarbault had seen was so small that the planet could never have sufficient mass to explain the excess rotation of Mercury's perihelion. But that was not necessarily an insurmountable problem: perhaps Vulcan, as Le Verrier called the new planet (after the Roman god of fire), was part of a whole belt of similar bodies. After all, more than 50 asteroids had been discovered between the orbits of Mars and Jupiter, so why could there not be a similar belt within the orbit of Mercury?

On January 2, 1860, Le Verrier gave a lecture on Vulcan at the French Academy of Sciences in Paris, and called on astronomers to look out for the new planet on July 18 of the same year. On that day there was to be a total solar eclipse, visible in northeast Spain and the extreme northeast of Algeria, when it should be possible to see the small planet close to the eclipsed Sun.

Vulcan was never found, but Le Verrier continued to believe in the planet's existence until his death in 1877. Others claimed to have actually seen the new planet: on April 4, 1875, a German astronomer saw a suspicious dot on the Sun, and two Americans thought they saw an unknown speck of light in the sky during the solar eclipse of July 29, 1878. That was a little less than a year before the birth of Albert Einstein, whose general theory of relativity ultimately offered the right explanation for the excess rotation of Mercury's orbit. Vulcan was no longer necessary, which was fortunate, as the observations were not particularly convincing. What Edmond Lescarbault saw in March 1859 will probably remain a mystery forever. After Einstein there was no need at all to continue to believe in the existence of a planet within Mercury's orbit.

Astronomy by Fighter Jet

So what does Dan Durda do in his F-18? Get a kick from flying a fighter jet, that's for sure. But he also scours the skies for celestial bodies within the orbit of Mercury – something which is practically impossible from the surface of the Earth. He doesn't expect to find a full-scale planet Vulcan – if that existed, it would have been discovered long ago. But he might find a population of small asteroids, no larger than a few tens of kilometers across: vulcanoids.

Durda conducted PhD research on cosmic collisions at the University of Florida in Gainesville. Not catastrophic encounters between planets, but smaller fry: asteroids, boulders, and lumps of rock flying around in space and smashing each other to smithereens. He hoped that his research would teach him more about the origin of dust particles in the solar system. But how do you study something like that? Certainly not by crashing two asteroids together – that is still beyond the realms of possibility. So you have to do tests in a laboratory, construct theoretical models, and simulate the evolution of the solar system using powerful computer codes.

His postdoctoral appointment at the University of Arizona in Tucson remains the high point of Dan Durda's career. It offered him more opportunities and more funds, without too many responsibilities. He worked with space scientists, simulating the impact that killed off the dinosaurs 65 million years ago, and finally got his pilot's license. What's more, it was a magnificent location and always beautiful weather – what more could a man want? Only a bombshell could have got Dan Durda to leave Tucson.

That bombshell hit in 1998, when Alan Stern of the Southwest Research Institute persuaded Durda to join him in Boulder. Stern, himself an avid pilot, offered him the only job in the world which allowed you to conduct planetary research from a fighter jet. It was an offer Durda couldn't possibly refuse. Together, he and Stern have now clocked up an impressive tally of flying hours.

Dan Durda (right) and Alan Stern in front of the fighter jet used to hunt for vulcanoids (Courtesy Dan Durda)

According to Durda, the fact that no vulcanoids have yet been discovered does not mean that they don't exist. Everywhere else where primordial matter from the formation of the solar system could be expected, it has always been found. And, unlike the zone between the orbits of Mercury and Venus, the area within Mercury's orbit is dynamically stable. Vulcanoids could be fragments of Mercury itself, left over from a severe impact. There can be nothing there larger than 60 kilometers across, otherwise Durda and Stern would have found them long ago. But there is a strong possibility that there are a few hundred 'hot asteroids' around 10 kilometers across orbiting the Sun. Not to mention many more smaller lumps of rock.

Earth-Grazers

Compared with giant planets like Uranus and Neptune, and even with a dwarf planet like Pluto, a cosmic rock 10 kilometers in diameter is of course not very impressive. But small asteroids in unusual orbits could actually have much greater significance for humanity than a large 'Planet X' in the outer reaches of the solar system. An object that remote has little effect here on Earth, but a rogue asteroid could mean the end of all human civilization.

No wonder then that planetary scientists have for many years tried to hunt down Earth-grazers: asteroids that might come dangerously close to the Earth's orbit, and which might even, at some time in the distant future, collide with our home planet. Sixty-five million years ago, the impact of a cosmic body some 10 kilometers across brought the age of the dinosaurs to an abrupt end. It doesn't bear thinking about that something like that might happen again in the near future.

The number of near-Earth asteroids larger than 1 kilometer across is estimated at around a thousand. Many of them have been tracked down and their orbits are sufficiently well known to assure us that we are not in danger, at least for the time being. But there is a large amount of unknown space garbage floating around in the inner solar system, and next week, a space rock could be found that is on course for a head-on collision with the Earth. That the Earth will be struck by an asteroid again at some time in the future is beyond doubt, the only question is when.

David Tholen (Photo archive of the Division for Planetary Sciences, AAS)

David Tholen of the University of Hawaii in Honolulu has plenty to say on this subject. He was one of the discoverers of Apophis, the 'most dangerous' Earth-grazer identified to date. Such a small object – Apophis is estimated to have a diameter of 300–400 meters – can be detected because it leaves a faint trail of light on a photo of the night sky. During the exposure, the asteroid moves with respect to the stars. If you measure that motion accurately, you can essentially calculate the object's path through the solar system.

That is, however, not an easy task. The method developed by Carl Friedrich Gauss in the nineteenth century to deduce the orbit of a celestial body from a series of positional measurements, does not work in many cases. Earth-grazers move so fast that there are usually not enough data points available. They leave a trail on a photograph, but by the time a new picture is taken, they have already disappeared from view. And it is even more difficult if they spend most of their time within the Earth's orbit, because then they are only visible in the twilight.

Tholen developed a new method that enabled near-Earth asteroids to be tracked. While Gauss focused on the orbital elements of an asteroid (the six numbers that determine the orbit of every planet), Tholen's method aimed at detecting three-dimensional positions and velocities. At first, this approach is less accurate, but on the basis of measurements taken at two different times, you can make a reasonable prediction of where the asteroid will be at some point in the future. If, on the basis of that prediction, it can then be photographed once again, its complete orbit can be pretty much determined.

Apophis

Apophis was discovered on June 19, 2004 and was long known simply as asteroid 2004 MN$_4$. It was clearly an Earth-grazer, but at first not enough was known about its orbit to say anything significant about the risk of it actually hitting the Earth. That was not possible in December, when the small asteroid was rediscovered in Australia, partly thanks to a rough estimate of its position by Tholen and his colleagues. And then the alarm bells started ringing.

2004 MN$_4$ proved to circle the Sun in a small elliptical orbit with an orbital period of 323.6 days – shorter than that of the Earth. Most of the time, the asteroid was within the orbit of the Earth, but twice during each revolution the two orbits crossed. If the Earth happened to be at that point at the same time as 2004 MN$_4$, we would be in serious trouble. It soon became clear that that would occur in April 2029. On Friday 13, to be exact.

At the end of December, after the asteroid's position had been measured several dozen times, the American space agency NASA announced that the chances of 2004 MN$_4$ actually colliding with the Earth on April 13, 2029 were almost 3%. Never before had humanity been faced with such a cosmic doom scenario. An asteroid a few hundred meters across might not be a threat to the survival of the human race, but the impact would cause a natural disaster of unprecedented proportions, and certainly many million deaths.

This doomsday prediction did not make the front pages. The tsunami which caused such havoc in Sri Lanka, Thailand, and Indonesia, understandably attracted much more attention than a natural disaster that might happen 25 years in the future. But for astronomers, 2004 MN_4 remained firmly at the top of the list of potentially hazardous asteroids, and they worked feverishly to collect more positional measurements and to refine their orbital predictions.

The cosmic projectile was not given an official minor planet number (99942) and a name until the summer of 2005. Apophis was the Greek name for Apep, the Egyptian god of destruction. It also happened to be the name of the evil alien in the television series *Stargate SG-1*, who wanted to destroy the Earth. However, the name soon lost its irony when more accurate calculations on Apophis' orbit, based on radar observation, showed that there would be no collision in April 2029. However, the asteroid would pass close enough to the Earth on that Friday 13 to be seen with the naked eye as a rapidly moving point of light in the night sky. As it made this 'low pass,' it would be closer than a number of communication satellites and its course would be markedly changed by the Earth's gravity.

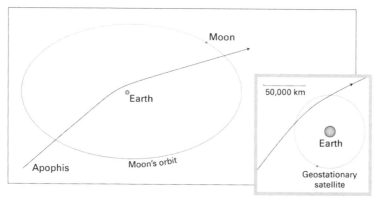

On April 13, 2029, Apophis will make an extremely close pass to the Earth (Wil Tirion)

Danger of Impact

And there lies the link with Dan Durda's vulcanoids. If there are lumps of rock several kilometers across flying around within Mercury's orbit, their orbits might also be changed if, for example, they pass too close to Mercury or Venus. And it is not inconceivable that they would then enter a wider orbit and pose a potential threat to the Earth. So anyone wanting to know the chances of a catastrophic cosmic impact occurring in the coming centuries needs to know what danger lurks in the innermost regions of the solar system.

If we do ever find an Earth-grazer that does prove to be on a collision course with the Earth, there might be something we can do about it. If the impact lies far enough in the future – say 20 or 30 years – there will be time to knock the asteroid slightly off course by using rocket motors or nuclear explosions, or by wrapping it in reflective material so that it reacts differently to the light from the Sun. Planetary scientists, astronauts, arms experts, and politicians regularly put their heads together to discuss these kinds of solutions, which sound like something straight out of science fiction books. The European Space Agency (ESA) even has plans to send an unmanned space probe to a small near-Earth asteroid to gain experience with the various scenarios.

The chance of a catastrophic impact can, however, never be completely ruled out. Perhaps new, advanced telescopes – like the Pan-STARRS asteroid hunter in Hawaii – will eventually chart all potentially dangerous space rocks in the inner solar system and conclude that the Earth is in no danger in the coming centuries. And perhaps Dan Durda and Alan Stern will come to the conclusion that vulcanoids do not exist and that there is no danger to be expected from the region inside Mercury's orbit. But even then, a cosmic Armageddon remains a possibility. Not from the inner solar system, but from the cold outer regions.

New comets are regularly discovered approaching the Earth at high speeds. It is not inconceivable that tomorrow, or next month, or in 10 years' time, one might be found that will be in the wrong place at just the wrong time. If that happens, there will be no time for us to take countermeasures, and mankind will be at the mercy of the cosmos.

Where do these potentially disastrous objects come from? From a flattened cloud beyond the orbit of Neptune, where billions of comets, tens of thousands of ice dwarfs, and a handful of frozen dwarf planets drift around, invisible from the Earth, on restless courses. Or from the Kuiper Belt, where Pluto lives with its larger brother Eris, which for a while was known as the tenth planet. It was the discovery and study of the Kuiper Belt that gave astronomers a completely new perspective on the early youth of the solar system, and which eventually led to the demotion of Pluto.

Chapter 15
The Kuiper Connection

Lucy Ann Kuiper was about 17 on that smoldering hot day in the mid-1960s when she walked through the centre of Tucson with her father. They had moved from the cold of Wisconsin a couple of years before, but she was still not used to the deadening heat of South Arizona. Fortunately the local bookstore had a good air-conditioning system. And what better way of spending time with your famous scientist father than to visit a bookstore?

Lucy loved reading and writing and had great admiration for her father. Yet she was a little startled when he took a biography of the physicist Enrico Fermi from the shelf and showed it to her. 'Look,' said the world-renowned planetary scientist, 'written by his wife. Perhaps one day, you'll write *my* biography.'

Gerard Kuiper at Kilauea, Hawaii, around 1968 (Dale Cruikshank)

G. Schilling, *The Hunt for Planet X*, DOI 10.1007/978-0-387-77805-1_15,
© Springer Science+Business Media, LLC 2009

That was typical of Gerard Kuiper. He was not known for his modesty. Most of his colleagues found him arrogant, even though they had to admit that he was also very good. Kuiper's student Dale Cruikshank put it very nicely: 'Kuiper had a very strong sense of the importance of his own work.' Cruikshank, now a famous infrared astronomer at NASA's Ames Research Center, had great respect for his teacher, who meant much more to him than his own father. But he knew better than anyone that the imposing scholar with his strange Dutch accent was not particularly easy to get on with.

But what else would you expect from the son of a country tailor from the north of the Netherlands? Gerrit Pieter Kuiper was born on December 7, 1905 in Tuitjenhorn, a small farming village between Bergen and Schagen. He spent his childhood jumping ditches, drawing maps ... and gazing at the stars. His father and grandfather gave young Gerrit a telescope as a present and in 1924, after he had completed the polytechnic, he went to Leiden to study astronomy. He was taught by the Austrian physicist Paul Ehrenfest and the great Dutch astronomer Jan Oort. And he was teased for his rural accent.

Luckily Kuiper was not the only provincial at Leiden Observatory. He spent a lot of time with Bart Bok, who came from the fishing village of Hoorn, not far from Tuitjenhorn. That was surprising, as they were so different they could have come from different planets. Gerrit was conceited and dour, while Bart was modest and sociable. A photograph from 1929, when the two recently graduated astronomers were doing their military service, shows them in typical poses: Kuiper at attention, jaw firm and resolute, cap pulled down over his brow; Bok with a wide smile, his cap at a jaunty angle, and one knee bent cheekily forward.

Clyde Tombaugh's discovery of Pluto, in the spring of 1930, must have made a deep impression on Kuiper. Under the close supervision of Willem de Sitter and Ejnar Hertzsprung he conducted statistical research into binary stars for his PhD, but his heart actually lay closer to home: with the Moon and the planets. Shortly after the discovery of the new planet – and by an American of the same age! – Kuiper wrote an article in *Hemel & Dampkring*, the monthly magazine of the Dutch Association for Meteorology and Astronomy, asking how much of Bode's Law could still be considered valid. His conclusion was: not much. Pluto did not fit in with the mysterious series of numbers which had so fascinated astronomers in the nineteenth century. A little later he wrote a series of articles on the planet Mars, which were also published in *Hemel & Dampkring*. One thing seemed clear: Gerrit Kuiper had set his sights on the solar system.

And on America. After being awarded his PhD in 1933 he left for Lick Observatory on Mount Hamilton in California and 2 years later he moved to the Harvard College Observatory, where he met his future wife Sarah Parker Fuller. Another 2 years later he became a naturalized American citizen, changed his first names to Gerard Peter and obtained an appointment at the University of Chicago's Yerkes Observatory in Williams Bay, Wisconsin. The imposing observatory, where the famed astronomer Otto Struve was at the helm, had the world's largest refractor telescope.

The 1-meter refractor of Yerkes Observatory (University of Chicago/Yerkes Observatory)

Beyond Neptune

Kuiper's career as a solar system scientist was a resounding success. In fact he was the founding father of American planetary science and has given his name to an asteroid, impact craters on the Moon, Mercury, and Mars, and a NASA aircraft converted into a flying observatory. And of course to the Kuiper Belt – the flattened cloud of icy celestial bodies beyond the orbit of Neptune, of which Pluto is one of the largest. Yet this was perhaps the one honor that Kuiper actually didn't deserve.

Doubts about Pluto's uniqueness were expressed as early as 1930. At the University of California in Berkeley, Armin Leuschner posited that the newly discovered object may have been a rogue asteroid or a giant comet. And if there is one such body, there may of course be more. Later that year Leuschner's younger colleague Frederick Leonard, of the University of California in Los Angeles, wrote these prophetic words in the *Leaflet of the Astronomical Society of the Pacific*: 'Is it not likely that in Pluto there has come to light the first of a series of ultra-Neptunian bodies the remaining members of which still await discovery, but which are destined eventually to be detected?'

It was, however, an Irish ex-soldier and retired amateur astronomer who first launched the theory in 1943 that there was a disk of small, comet-like objects beyond the orbit of Neptune. At the age of 63, Kenneth Edgeworth wrote an article about his theory in the *Journal of the British Astronomical Society*, a publication for amateur astronomers. Six years later, he outlined his ideas in much greater detail in the scholarly *Monthly Notices of the Royal Astronomical Society* and in 1961, at the age of 81, he published a book entitled *The Earth, the Planets and the Stars*, in which he also presented the comet-disk theory.

Edgeworth's reasoning was actually very simple. He assumed that, long ago when the Sun was very young, the planets coagulated together in a flat, rotating disk of gas and dust. This process of accretion had naturally started with the formation of small fragments and condensations. Only later did they combine to form larger bodies. But according to Edgeworth it was improbable that the disk had a sharp outer edge. It was more logical that it tapered off gradually. If that were true, there must be a large number of smaller objects beyond Neptune's orbit which had never come together to form full-fledged planets.

In the mid-1940s, little was known with certainty about the formation of the solar system, but many scientists toyed with the idea of a turbulent disk of matter in which small particles eventually grew into larger bodies. In 1946, for example, Hendrik Petrus Berlage, the son of the famous Dutch architect, submitted a long article about this theory to *The Astrophysical Journal*, the editors of which had their offices at Yerkes Observatory. It was never published but it must have aroused the interest of Gerard Kuiper there in Williams Bay; and Kuiper is sure to have seen Edgeworth's publication in the *Monthly Notices* a couple of years later.

The Kuiper Belt

It is therefore remarkable to say the least that neither Berlage nor Edgeworth appear in the bibliography of an extensive article entitled *On the Origin of the Solar System* which Kuiper wrote in the early 1950s. The article was published in 1951 as Chapter 8 of a thick report on the symposium marking Yerkes Observatory's 50th anniversary. The book was edited by Allen Hynek of Ohio State University, who was to later become famous for his research into

UFOs and even had a guest role in the film *Close Encounters of the Third Kind* in 1977.

Kuiper's article did, however, mention his old teacher at Leiden, Jan Oort, although Kuiper denied the validity of Oort's comet cloud theory. In 1948, Oort had predicted the existence of an enormous cloud of frozen comets at an extreme distance from the Sun. He thought that these cosmic snowballs could long ago have been formed in the asteroid belt and had some time later been catapulted into space by Jupiter's gravitational force. But, according to Kuiper, the comets in the Oort Cloud came from a ring of matter beyond the orbit of Neptune and it had been Pluto that had caused their expulsion.

The same theory of the origin of the solar system that described by Berlage, and the same argument for material beyond the orbit of Neptune that Edgeworth had proposed, and no reference to either – is that coincidence or scientific plagiarism? Perhaps, more than anything, it was typical Kuiper: his own ideas were always just a little better and more complete than those of others. Edgeworth was not even a professional astronomer and Kuiper only included references to other people's publications if he knew them personally.

However it happened, the extended disk of icy bodies in the outer regions of the solar system – ranging from small comets to dwarf planets more than 2,000 kilometers across – has been called the Kuiper Belt since the early 1990s, with reference to Kuiper's chapter in Hynek's book. To the great displeasure of Irish astronomers, who would rather have seen it called the Edgeworth Disk, and to the annoyance of several Americans who have read Kuiper's article again with a critical eye and reached the conclusion that the great planetary expert had predicted many things, but the existence of the belt that carries his name was not one of them.

So how did it happen? Like Edgeworth, Kuiper believed that in the early days of the solar system, small aggregates of icy matter must have formed beyond Neptune's orbit, and perhaps a handful of larger objects like Pluto and Eris. But all of those small comets should have been catapulted into outer space long ago by the gravitational force of Pluto – a planet which, according to Kuiper, was nearly half the size of the Earth. If his theory were right, the outer regions of the solar system would be empty by now. In other words, there would be no Kuiper Belt.

Even Gerard Kuiper would probably have been very surprised that the belt was named after him. In his chapter on the formation of the solar system he only devoted a couple of paragraphs to the origin and evolution of cometary orbits, and he did not pursue the matter later in his career. When he died at the end of 1973, none of his colleagues or former students (including the famous popularizer Carl Sagan) mentioned the Belt in their obituaries. Kuiper's violently expelled cometary belt was a partly borrowed footnote, an erroneous sidestep in his scientific career.

Enormous Legacy

A career which otherwise left little to be desired. When, during the Second World War, Kuiper was set to work at the Radio Research Laboratory of Harvard University, as a renowned astronomer he was given time off in the winter of 1943–1944 to pursue his own research. That winter the giant planets Jupiter, Saturn, Uranus, and Neptune were all clearly visible, and Kuiper observed the large moons of the four planets in detail with, for the time, sensitive spectroscopes to determine whether they might have an atmosphere.

He made his observations of course with the 2.1-meter Struve Telescope at the McDonald Observatory on Mount Locke, in the Davis Mountains in western Texas. At the end of the 1930s, Kuiper had been closely involved in the building of this observatory, which had been made possible by a donation by the Texan banker William Johnson McDonald to the University of Texas in Austin. That university, however, had no astronomy faculty, so they asked the University of Chicago for assistance from the astronomers at Yerkes Observatory.

At the time the Struve Telescope was the third largest in the world and in 1944 Kuiper used it to make an astonishing discovery: the large Saturnian moon Titan has an atmosphere containing methane gas. This was the first time that a planetary satellite had been found to have an atmosphere. A few years later, shortly after the birth of his daughter Lucy Ann, Kuiper discovered small, unknown satellites on images of the distant giant planets Uranus and Neptune. He initially wanted to call the new Uranian moon (discovered in 1948) Lucy, but he ultimately settled for the name Miranda. The Neptunian moon that he found a year later was named Nereid. In between times, Kuiper also discovered carbon dioxide in the atmosphere of Mars and showed that the particles in the rings of Saturn are covered with a layer of ice.

Lucy was the apple of her father's eye – much more than her older brother Paul. Kuiper read her stories before she went to bed, woke her up in the middle of the night to look at the Pole Star or a pretty planetary conjunction, and once even gave her a necklace with real meteorite stones for her birthday. She thought it was fantastically exciting having a father who (from 1947 to 1949 and again from 1957 to 1960) was the director of two observatories. There were always interesting and handsome young men around the place – new first-year students – and she remembers as if it were yesterday the visit by a delegation of astronomers from the Soviet Union, at a time when Communist-hunter Joseph McCarthy was in the news almost daily: they had to be Russian spies!

In the early 1960s, the family moved to Tucson. It was not entirely their own decision: Kuiper was not very popular with his colleagues in Williams Bay, who saw little point in all that interest in planets and satellites. In Arizona, together with a small group of colleagues and students, he set up the Lunar and Planetary Laboratory. Later he founded the Catalina Observatory, near Tucson, was involved in building the Mauna Kea Observatory in Hawaii, took a great

Gerard Kuiper, observing at Catalina Observatory, around 1969 (Dale Cruikshank)

interest in the origins of the Moon, and helped NASA to select landing locations for the Apollo missions.

Working hard, at night too, looking through the eyepiece of the telescope after short catnaps of 2 or 3 hours, and never being satisfied with anything but the best – that was Gerard Kuiper all over. It was therefore no wonder things went wrong so quickly after he retired. At the end of December 1973, when he had just stepped down as director of the Lunar and Planetary Laboratory, Kuiper and his wife Sarah were invited to Mexico City, where a documentary was being made about the Mayas and the Aztecs, who were also very knowledgeable about astronomy. But the thin air and walking up and down the many steps of the temples of the Sun and Moon for the cameras proved fatal. Kuiper died suddenly on Christmas Eve after suffering a heart attack in his hotel.

The fact that his son Paul (named after Kuiper's mentor Ehrenfest) was interested in astrology must have been a thorn in Kuiper's side; perhaps that was why he was such a dominant father figure for his student Dale Cruikshank. And he must have been upset when his daughter Lucy Ann later joined a spiritual community where she 'discovered' that her true first name was Sylvia. But in the field of solar system science, Gerard Kuiper left behind an enormous legacy. He may not have been the first to have thought about icy bodies in the outer regions of the solar system beyond Neptune, but in a certain sense it is fitting and appropriate that the wide belt that encloses the entire solar system is named after him.

And that biography? Lucy Ann, alias Sylvia, never wrote it. In fact, until now, no one has dared to tackle the subject. It is too comprehensive, too elusive and, perhaps also just a little too sensitive. Anyone who takes on the job, can hardly avoid portraying Kuiper not only as a single-minded genius, but also as an arrogant know-it-all. In Tuitjenhorn they have found a simpler solution: they have erected a statue to Gerrit Pieter and named a street after him. No one can argue with that.

Chapter 16
Comet Puzzles

Martin Duncan can watch them for hours. Little dots wriggling around on the computer screen. In the middle is the Sun. The four larger dots circling it are the giant planets: Jupiter, Saturn, Uranus, and Neptune. All of the others, smaller ones beyond the orbit of Neptune, are comets. They move around the center in a slow procession, like Muslim pilgrims around the Kaaba during the Hajj. But every now and again, one of them will fall toward the center, its orbit disturbed for some reason or another. Once a comet comes adrift in this way, sooner or later it will cross the orbit of one of the giant planets and receive another jolt from its gravity. Eventually it will end up in a small, elliptical orbit, which will take it through the inner solar system on each circuit. It will then pass close to the Earth, where people will gaze in wonder at the impressive sight of a 'star with a tail.'

Duncan himself did not become an astronomer after seeing a spectacular comet in the night sky. Little Martin was born in London and grew up in Montreal, Canada – and a comet must be extremely bright if you are to see it from within a large city. He has never been a real stargazer: he can't recognize the constellations and there's no point in asking him where Saturn is in the night sky. And the few times that he has stood and waited to see a comet have usually come to nothing. The long-awaited return of Halley's Comet in 1985 and 1986, for example, which was announced with such a wave of publicity, was a great disappointment. The comet was hardly visible with the naked eye from the northern hemisphere, and even through a telescope it was hardly impressive.

Comets are usually named after the person who discovers them, but Halley's Comet is an exception. The English astronomer Edmund Halley, who studied the comet in 1682, discovered that its orbit was suspiciously similar to those of two previous comets observed in 1531 and 1607. Halley concluded that they were one and the same object, which passed close to the Earth every 76 years. The comet was indeed seen again in December 1758 and immediately named after Halley, who by then had been dead for 16 years.

With their spectacular appearance, comets are the 'glamour boys' of the solar system. Normally they would pass by unnoticed, but once they dress themselves up with glowing heads and wide, fan-shaped tails, they can no longer

G. Schilling, *The Hunt for Planet X*, DOI 10.1007/978-0-387-77805-1_16,
© Springer Science+Business Media, LLC 2009

be ignored and demand our full attention. Generals and rulers have trembled at the appearance of an ominous comet, slicing through the firmament like a saber of light, and even in 1997, when Comet Hale–Bopp was visible in the sky, superstition was rife.

But this whole visual display is a glorious deception. A comet is actually nothing more than an irregular-shaped lump of ice and rubble a few kilometers across, a dirty, frozen snowball, much too small to be seen from the Earth. But once this cometary nucleus approaches the Sun, the ice starts to evaporate, creating geysers and fountains of gas, and releasing dust particles and grit. The comet is enveloped by an extended, tenuous cloud of gas and dust particles, which form the head (or coma), and the Sun blows all the matter away in a long tail, which can sometimes be tens of millions of kilometers in length.

Comets therefore by no means live forever. Every time their elongated elliptical orbits bring them close to the Sun, they lose thousands of tons of ice and dust. The porous nucleus becomes smaller and smaller until eventually the comet falls apart. Clearly comets do not always end up in these small elliptical orbits, otherwise they would all have disappeared long ago – if you see a popsicle lying on the table of a beachside café and melting in the Sun you can be pretty sure it hasn't been there for long either. But where do comets come from? Who keeps putting new popsicles out in the Sun?

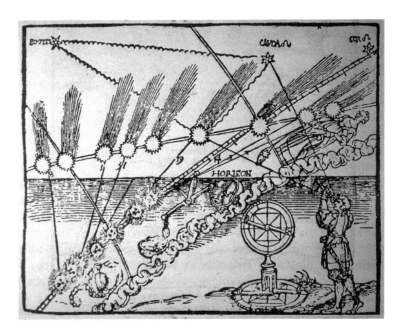

Ancient woodcut of cometary observations, showing that a comet's tail always points away from the Sun

The Oort Cloud

Shortly after the Second World War, PhD student Adriaan van Woerkom was studying the evolution of cometary orbits at Leiden University. He focused mainly on the gravitational influence of the planets. The massive giant planets can capture whole families of comets. We now know of several hundred comets which, at the furthest point of their elliptical orbits, are almost equally as far from the Sun as Jupiter. They probably once had much wider orbits and were drawn into these smaller, shorter orbits by Jupiter's gravitational pull.

Jan Oort (Courtesy Stichting De Koepel)

Van Woerkom's research attracted the attention of Leiden professor Jan Oort. Oort was, however, much more interested in long-period comets, which take more than 200 years to orbit the Sun. Some of these comets have such an elongated orbit that it must take them tens of thousands of years to complete a circuit. We therefore see such comets only once: they emerge from the dark void, make a spectacular hairpin turn around the Sun, and then disappear again into the far distance. And they come from all possible directions.

In a ground-breaking article in the *Bulletin of the Astronomical Institutes of the Netherlands* in 1950 Oort explained that there was only one solution for the origin of the long-period comets. Our Sun must be surrounded at a great distance by a gigantic, spherical cloud of cometary nuclei, something that had been suggested before by the Estonian astronomer Ernst Öpik. Although these nuclei are on average billions of kilometers apart, they sometimes pass close to each other, disturbing their orbits. That can lead to a comet falling into the inner solar system. From the number of long-period comets that appear in the sky each century, Oort calculated that the cloud may contain some 10 billion cometary nuclei.

That was quite a feat: from the data on the orbits of a few dozen long-period comets Oort concluded that there must be a cloud with a diameter of a few light-years and containing more comets than there are stars in the Milky Way. But his calculations were watertight and, although the Oort Cloud has never actually been seen, no one doubts that it exists. So where did the comet cloud come from? Oort had an answer to that, too: the dust and rubble in the cloud probably originated in the asteroid belt and were catapulted into space by Jupiter's gravity.

Oort's former student Gerrit Kuiper suggested a year later that the comets in the Oort Cloud did not come from the inner solar system, but from its outer regions, beyond the orbit of Neptune. But, like his teacher, Kuiper believed that all the comets that could now be observed in the sky had their origin in the colossal Oort Cloud. An icy cometary nucleus would be ejected from its orbit by another nucleus or perhaps by the gravitational influence of a passing star, fall toward the Sun and make its appearance as a long-period comet with an orbital period of tens of thousands of years. During one of these circuits, it might happen to pass a giant planet and be forced into a short-period orbit and then, over a period of a few thousand years, fully disintegrate.

It was a nice story, but we know now that not much of it is true. In the mid-twentieth century, however, there was no convincing way of confirming or refuting the validity of such a theory. It sounded reasonable and, as an astronomer, you had to make do with it. Until 1990, that is, when Martin Duncan and his colleagues found the correct explanation for the origin of the short-period comets. Their revolutionary computer simulations showed that these comets do not come from the Oort Cloud, but from another 'comet reservoir' just beyond the orbit of Neptune: the Kuiper Belt.

Roving Comets

Martin Duncan was born in the year when Oort published his article on the comet cloud. He was a real thinker; even as a child he would continually ask himself, what would happen if. . .? He preferred to ask questions that couldn't be answered experimentally, but only through theoretical thinking. Martin was therefore never a practicing astronomer and he never had his own telescope. After studying astronomy at McGill University in Montreal he threw himself (figuratively speaking) heart and soul into black holes and, in 1980, was

Martin Duncan (Queen's University)

awarded his PhD at the University of Texas in Austin for his theoretical research into supermassive black holes in the cores of galaxies.

Black holes are completely different from roving comets. They are greedy cosmic gluttons – bizarre warps in space-time, ruled by an incomprehensible mix of quantum physics and relativity. How do you get from there to the indolent world of slowly orbiting celestial bodies in our own solar system? It is all the fault of Nemesis. In the mid-1980s, the idea that the Death Star causes the Earth to be bombarded every 26 million years with a barrage of comets was popular and exciting. Duncan, by then working at the University of Toronto, wondered what would happen if this small companion to the Sun were to move through the Oort Cloud.

There were, incidentally, striking similarities with his work on black holes. Nemesis disturbs the motion of neighboring comets just as a massive black hole disturbs the motion of neighboring stars. And in both cases there is a danger zone: if a star comes close to a black hole, it is ripped to pieces and swallowed up; and if a comet comes too close to Nemesis, it is catapulted out of its orbit. In general terms, both phenomena are explained by the same math, the same formulae.

Nemesis was soon pretty much forgotten – according to Duncan, Richard Muller is just about the only person left in the world who still believes in it – but Duncan's interest in the dynamics of the solar system is as strong as ever. This was largely due to his stimulating collaboration with Scott Tremaine, who set up the Canadian Institute for Theoretical Astrophysics (CITA) at the University of Toronto in 1985. Thomas Quinn was hired as a postdoc and the three of them sank their teeth into computer simulations of the evolution of cometary orbits.

Wriggling Dots

When Duncan, Quinn, and Tremaine tried their hands at solving this astronomical problem with computers in the mid-1980s, they were of course not the first. The Uruguayan astronomer Julio Fernández of the University of the Republic in Montevideo had already tried his luck at the end of the 1970s. Fernández, a great lover of science fiction who had become interested in the cosmos when Sputnik was launched in 1957, taught himself the computer programming language Fortran and recalls that, for each calculation, he had to walk back and forth with piles and piles of punch cards.

Julio Fernández (Courtesy Julio Fernández)

For almost 30 years Oort's 'sounds-good' theory had been the standard explanation for the origin of short-period comets. They supposedly came from the Oort Cloud and for some reason or another ended up in small orbits. But Fernández was not convinced. The orbits of the long-period comets, which almost certainly come from the Oort Cloud, have widely varying orientations. They are often at a steep angle to the plane in which the planets move around the Sun, and about half of them orbit the Sun 'in the wrong direction,' opposite to that of the planets. The short-period comets on the other hand are usually at a shallow angle to the orbital plane and orbit in the 'right' direction.

Fernández showed that the capricious long-period comets cannot as easily be 'captured' in small, 'regular' orbits – the chances of that happening are simply very small. Yet, since we know that an extremely large number of short-period comets exist, and because they all have a relatively limited life span, there must be a constant supply of new ones – just think of the popsicles in the Sun. So do they perhaps come from somewhere else after all?

Who knows? American comet expert Fred Whipple wrote about this puzzle in the mid-1960s. He suggested that there may be a large reservoir of comets beyond the orbit of Neptune – a wide ring of icy fragments, remnants of the time when the solar system was formed. With this idea, Whipple was following in the footsteps of Kenneth Edgeworth and Gerard Kuiper, but in contrast to Kuiper (who believed that the small fragments had been ejected from the solar system long ago) Whipple thought it possible that the reservoir was still being replenished. And if such a flat ring really is a remnant of the formation of the solar system, it is not so surprising that the short-period comets should move in orderly orbits, in more or less the same plane as the planets and in the normal direction.

Another good idea, but how does a comet get from Whipple's reservoir to the inner solar system? With the limited computer capacity available to him at the end of the 1970s, Fernández tried to solve that puzzle. Perhaps, he reasoned, the ring of comets also contained larger bodies, many hundreds of kilometers across – at such a great distance from the Earth they would easily escape the probing eye of even the biggest telescopes. These larger bodies could disturb the orbits of the smaller comets. And, some of the latter might then fall under the influence of Neptune and experience such a radical change of orbit that they would move close to Uranus. Later, perhaps, the same could happen with Saturn and Jupiter. In this way, as his computer simulations showed, a comet could find itself in an ever smaller orbit. It would in effect be like a ping–pong ball bouncing down a staircase, with the four giant planets acting as the stairs. Eventually the comet would end up in an orbit entirely within that of Jupiter.

Simulations

Fernández published his results in 1980. His article went largely unnoticed at the time, however, perhaps because planetary scientists had their hands full with the spectacular results sent back by the American Pioneer and Voyager probes.

But 5 years later, it was read with great interest by Scott Tremaine, who then studied older publications by Whipple and Kuiper. Together with Martin Duncan and Thomas Quinn, Tremaine decided to take another quick look at Fernández's calculations. Or rather a long, hard look. The more powerful computers available at CITA allowed them to do the same job with much greater precision and reliability.

Those computers, the largest cost item on the research project's budget at the time, were downright primitive compared to today's mainframes. Perhaps a thousand times slower, Duncan thinks. The specially written software, which the three researchers had spent months working on, had to be tailored to the computers' relatively limited calculating capacity. To speed up the time-consuming calculations, they made the giant planets much more massive in their simulations than they were in reality. This meant that their gravitational influence was much stronger, so that the orbits of the simulated comets changed more quickly.

Was that not a little risky? How do you know for certain that the results of your calculations genuinely reflect reality? According to Duncan, the distortions were not that serious, though it was of course important to be cautious and remember that it was a only a computer simulation and not real life. 'What would happen if. . .?' is a dangerous question with these kinds of simulations. It is easy – and sometimes very tempting – to choose the input of your program so as to ensure that you get exactly the output you expect.

That output, by the way, consisted of unbelievably long series of figures: tables showing masses, coordinates, and velocity vectors for every moment in the simulation. And that meant developing separate software to present the whole mishmash of numbers in a way that would appeal to the imagination: as a sort of film in which you could see the solar system evolving on fast forward. After the computers in Toronto had spent months crunching away at the figures, there was a visible result: a collection of wriggling dots on Duncan's monitor.

It was a result to be proud of. At the end of the 1980s, Duncan, Quinn, and Tremaine wrote three articles about it in leading scientific journals: the first in 1987 on the formation of the Oort Cloud (1987) and two, in 1988 and 1990, on the origins of the short-period comets. According to the authors, there could no longer be any doubt about the latter. The detailed computer simulations showed that the comets in the inner solar system did not originate in the Oort Cloud, but in an as yet undiscovered reservoir beyond the orbit of Neptune. In the title of their article they called the reservoir the 'Kuiper Belt.'

The articles hit the astronomical world like a bombshell. What made them so different to Edgeworth's virtually forgotten publications, Kuiper's almost marginal notes, Whipple's fantasies, and Fernández's practically unnoticed article? Perhaps Duncan and his colleagues had a better feeling for public relations, or the time was ripe for a serious answer to an age-old question. Or perhaps this time, the result was based on a thorough scientific analysis – the indisputable outcome of a theoretical experiment instead of a hypothesis that sounded reasonable but for which there was no proof.

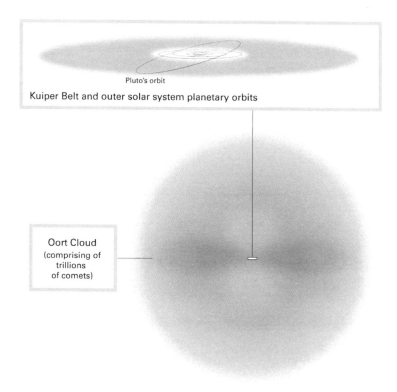

Pluto's orbit

Kuiper Belt and outer solar system planetary orbits

Oort Cloud
(comprising of
trillions
of comets)

Schematic drawing of the Oort Cloud and the Kuiper Belt (Wil Tirion)

Duncan received invitations for lectures and presentations, the term 'Kuiper Belt' achieved immediate popularity and, in 1989, the hunt for the comet belt was opened. If, in addition to small cometary nuclei, there were also larger icy bodies beyond the orbit of Neptune, the latest generation of electronic cameras might just be able to detect them. The discovery of the first Kuiper Belt objects would surely not be long in coming and would be a fantastic confirmation of the theoretical work of the Canadian researchers.

No one seemed to realize that the first Kuiper Belt object had already been discovered 60 years earlier, by Clyde Tombaugh in 1930. Pluto – wasn't that a planet? It was then, but it wouldn't be for much longer.

Chapter 17
Smiley

'I'm coming to Hawaii and I'm going to get you.' Dave Jewitt couldn't believe his eyes, but that was what it said. It was a threatening email. To an astronomer. You don't imagine something like that. The name of the sender meant nothing to him, but this unknown American seemed determined to jump on a plane and come over and demand satisfaction. And why? Because his son Billy was so upset by the news that Pluto may no longer be a planet. The enraged parent blamed Jewitt and was out for revenge.

That was in 1995. Jewitt never received a visit from Billy's father or a follow-up to that first email, but even for a level-headed Englishman he was a bit ruffled. As an astronomer you don't expect to find your personal safety under threat because you have made a spectacular discovery on the periphery of the solar system. Clearly, many people can't accept things they learned long ago at school being called into question. Things like the status of Pluto, for example. Jewitt has never made any secret of his opinions on that thorny issue. As far as he is concerned, ever since he discovered the first ice dwarfs in the Kuiper Belt, there is no way in the world that you can reasonably continue to call Pluto a planet. Sorry, Billy.

Jewitt knows from experience that the cosmos can have a great impact on young children. Dave was seven when he saw a shower of shooting stars in the night sky in November 1965. It aroused an interest in the universe that has never faded. His grandparents gave him a telescope and, after secondary school, he went to study astronomy at the University of London, where he was awarded his Bachelor's degree in 1979 at the age of 21. But England is by no means the best place for astronomical observations, so Jewitt moved to Pasadena to conduct planetary research for his Masters degree at the California Institute of Technology (Caltech).

Jewitt arrived at just the right time. That summer the American space probe Voyager 2 flew past the giant planet Jupiter. Together with his study supervisor Edward Danielson and Stephen Synnott from the nearby Jet Propulsion Laboratory (JPL), Jewitt studied the Voyager images of the tenuous ring of dust around Jupiter. On one of the photos he discovered a small moon, which was later given the name Adrastea. Not a bad start for someone who had not yet even begun his PhD research.

G. Schilling, *The Hunt for Planet X*, DOI 10.1007/978-0-387-77805-1_17,
© Springer Science+Business Media, LLC 2009

That research focused on observations of comets, for which Jewitt frequently used the Caltech telescope park on Palomar Mountain. He even sat in the old-fashioned observing cage of the 5-meter Hale Telescope – high up at the front end of the telescope tube – observing the Moon and Jupiter. And on October 16, 1982, again together with Danielson, he was the first to rediscover Halley's Comet, on an electronic image made with the Hale Telescope. The famed comet, with its orbital period of 76 years, would pass through the inner solar system 3.5 years later, but Jewitt and Danielson found it while it was still far beyond the orbit of Saturn, an extremely faint speck of light in the constellation Canis Minor.

Unfortunately, Jewitt no longer had access to the Caltech telescopes after he became an associate professor at the Massachusetts Institute of Technology (MIT) in Boston in 1983. That was a pity, because he had become increasingly interested in the outer regions of the solar system and why they were so conspicuously empty. With the exception of Pluto there was nothing discernible beyond the orbit of Neptune; since early 1979, even Pluto's eccentric orbit had taken it closer to the Sun than Neptune – it would not be the most distant planet again until 1999. A targeted search with a large telescope should reveal more. Jewitt had not been very impressed by the lone quest conducted by Charles Kowal: that had resulted only in the strange centaur Chiron between the orbits of Saturn and Uranus. He was sure he could do better than that.

Unlike Kowal, Jewitt did not intend to search the entire zodiac. If there were small bodies orbiting the Sun beyond Neptune – aggregations dating back to the origins of the solar system which had never come together to form a full-fledged planet – there must be so many of them that you should be able to find them in any direction, as long as you made long exposures with a sensitive telescope. Jewitt started his search for 'trans-Neptunian objects' in 1985 using an old Schmidt Telescope at Kitt Peak National Observatory in Arizona.

It was laborious work. Everything had to be done by hand; the photographic plates had to be developed at night in the observatory's darkroom and, like Clyde Tombaugh and Charles Kowal, Jewitt spent many hours at the blink comparator – 35 years after the discovery of Pluto there had been little in the way technological advances on that front. The most frustrating thing, of course, was that he found nothing, despite the fact that he eventually searched an area of the night sky 500 times larger than the Full Moon.

Uncertainties

From early 1987 Jewitt was assisted in his lonesome task by his student Jane Luu, who needed a subject for her thesis and allowed her study supervisor to persuade her to work with him. It was a risky choice, because there was no

David Jewitt and Jane Luu, who discovered the first Kuiper Belt Object (Courtesy David Jewitt)

certainty that the search would be successful. But Luu was not someone who would easily be deterred by uncertainty – her whole life had been a series of surprises and uncertainties.

Luu Le Hang was born in South Vietnam in 1963, during the Vietnam war. Her father worked as a translator for the American army. She was a small girl of 11 when the North Vietnamese army took Saigon in April 1975. With her parents, her older sister, and two younger brothers, Le Hang was evacuated by the Americans and the family spent a month at an enormous refugee camp on a navy base in California. In the United States, Le Hang changed her name to Jane and, after a short stay in Kentucky with a younger sister of her mother's who was married to an American soldier, she returned to California where she sailed through high school in Ventura with flying colors.

But, in all honesty, Jane Luu had no idea at all what she wanted to do later. Through friends, she ended up at Stanford University, where she studied physics, and in 1984 – again by coincidence – she got a job at JPL, NASA's control centre for unmanned space research. At JPL she mainly did programming work for the Deep Space Network, a network of radio dishes which received the signals from space probes. But she became fascinated by the wonderful images that hung everywhere on the notice boards at JPL, such as the magnificent close-ups of Jupiter's moons and the rings of Saturn sent back by the Voyagers in the late 1970s and early 1980s.

And that led to her decision, in the fall of 1986, to do a Masters degree in planetary science at MIT in Boston. It was a lot more interesting than conducting physics experiments with heat exchangers and vacuum pumps in dark, concrete basements. On the unknown – and therefore inviting – East Coast, Luu started her research into the possible links between asteroids and comets – a subject which hitherto had received very little attention.

Very shortly after Jane Luu came to assist him, Dave Jewitt had had his fill of the Schmidt Telescope's photographic plates. The emulsion was apparently not sensitive enough to detect small, faint objects beyond the orbit of Neptune. He refused to believe that such objects simply did not exist. Perhaps the new Charge Coupled Device (CCD) technology would produce better results? The electronic camera he had used to rediscover Halley's Comet in 1982 was already equipped with a CCD device – a detector that had been developed for military applications, but which was now also used in astronomy.

Today, large CCD detectors with a few million pixels are used in consumer digital cameras, but in the mid-1980s, this technology was still in its infancy. One of Jewitt's colleagues at MIT, George Ricker, had developed and built an electronic camera with 70,000 pixels, a lot for the time. The biggest disadvantage of these CCD cameras was that, because the detector was so small, they had a very small field of view. On the other hand, a CCD is much more sensitive than a photographic plate with the same exposure time and the image is immediately available in digital form for further computer processing and electronic analysis.

In the spring of 1987 Jewitt and Luu mounted Ricker's camera in the focal plane of the 1.3-meter McGraw-Hill Telescope at Kitt Peak in Arizona and recorded 70 images in the course of 1 week. They covered an area of the night sky about one and a half times as large as the Full Moon. This was a thousand times smaller than the area covered earlier by the Schmidt Telescope, but because of the higher sensitivity, objects could be detected that were a hundred times fainter than those which could be seen with the Schmidt. Unfortunately, that project too produced no results: the images showed countless asteroids in the inner solar system, but nothing beyond the orbit of Neptune.

They were disappointed, of course, but Dave Jewitt saw no reason to give up. He was determined to be the first to find an object in the Kuiper Belt (the name had been introduced at the end of the 1980s by Scott Tremaine and his colleagues in Canada) and with better telescopes and larger CCD cameras, it had to be possible. It was this belief that led Jewitt to move in 1988 to the astronomical institute at the University of Hawaii in Manoa, a suburb of Honolulu. Astronomers at the institute had access to all the telescopes at Mauna Kea Observatory on nearby Big Island. If they didn't find anything here, they wouldn't find it anywhere.

White Mountain

Mauna Kea ('white mountain') is a 4,200-meter high, dormant volcano. The highest point in the Pacific, it is a landscape of capricious rock formations and multicolored volcanic boulders. The air is so thin that many people suffer from headaches, dizziness, shortage of breath, and loss of concentration. Yet, in recent decades, the world's largest observatory has been built on this high peak. It boasts 13 large telescopes, including the two identical Keck Telescopes (each with a mirror 10 meters in diameter) and the Japanese 8.3-meter Subaru Telescope. At 3,000 meters, alongside the winding road to the summit, is Hale Pohaku ('house of stone'), where astronomers can sleep and eat in slightly more comfortable surroundings. If possible, they spend the night at Hale Pohaku before visiting the observatory to allow their bodies to acclimatize to the low oxygen content in the air and thereby reduce the risk of altitude sickness.

When Jewitt came to Hawaii, the giant telescopes on Mauna Kea had not yet been built and the road to the summit was still largely unpaved. As remote control had not yet been developed, once a month (around the New Moon, when the nights are very dark) he and Jane Luu would spend a week on the mountain, where life is difficult and unpleasant, but the view is magnificent and the quality of the observations optimal. The University of Hawaii's 2.2-meter telescope which Jewitt and Luu used may have been quite a lot smaller than the Hale Telescope on Palomar Mountain but, because of its shorter focal length, the much higher optical quality of the mirror, and the unique location, you could observe much fainter objects. In addition, there were now CCDs available with 640,000 – and later even more than a million – pixels. In short: it was only a matter of time before the first object in the Kuiper Belt would be found.

Some of the telescopes on Mauna Kea Observatory. The University of Hawaii's 2.2-meter telescope is the fourth from the left (Govert Schilling)

The University of Hawaii's 2.2-meter telescope with which 1992 QB$_1$ was discovered (Govert Schilling)

There was by now good reason not to waste any time; there were competitors on the horizon. Martin Duncan, whose computer simulations had convinced a lot of astronomers of the existence of the Kuiper Belt, had started a search himself at the end of 1988, together with Hal Levison of the United States Naval Observatory in Flagstaff. At the end of 1990 at the University of Texas, husband-and-wife team Bill and Anita Cochran had started to search for ice dwarfs in the outer regions of the solar system. And when, on January 9, 1992, David Rabinowitz discovered the second centaur with the University of Arizona's Spacewatch Telescope – a large, red brother for Chiron in a highly inclined, eccentric orbit – it was obvious that surprising discoveries were possible with the new, sensitive CCD cameras.

In the summer of 1992 a new CCD camera with more than 4 million pixels became operational on the 2.2-meter telescope at Mauna Kea. That was a great improvement on the previous detector, which had only a million pixels. This increased the speed of the search fourfold. Naturally, expectations were high – would anything finally be found with the new camera? On Friday, August 28, it was New Moon and on the following day a new session of observations began lasting five nights. Yet again, Jewitt and Luu flew from Honolulu to Hilo, drove along the winding Saddle Road to Hale Pohaku and, together with their night assistant, made the final leg of the journey to the mountain top, with thick pullovers against the cold in the control room and books and music to help pass the time.

Lucky Strike

On the second night, Sunday, August 30, they struck lucky. Earlier that evening four images had been recorded of an area of the night sky in the constellation of Pisces, and while Luu was chatting to the night assistant, Jewitt had started blinking the first two photographs. That is a little easier and more accurate with digital images than with photographic plates. You no longer have to go into the darkroom and you don't need a blink comparator; you can now do it all on your computer monitor. The little point of light jumping on the screen was difficult to miss. 'Jane, come and take a look,' Jewitt said, 'I think we've got a bite.' The moving point of light was also visible on the third and fourth images of the same part of the night sky. They soon decided to depart from the planned observing program and focus all their attention on the new body. Could it be a nearby asteroid that is moving more or less toward us or away from us and therefore seems to move only slightly across the sky? No, because then there would be an extra change of position during the course of the night caused by the rotation of the Earth. The conclusion was clear: this object was located beyond the orbit of Neptune, perhaps 8 billion kilometers from the Sun. To be so clearly visible at such a distance it must be at least a few hundred kilometers in diameter, assuming that it reflects as much sunlight as Pluto. And if you realize that only a minute piece of the sky had been observed at this sensitivity, you could predict that there must be thousands of similar objects circling around in the cold, dark outer regions of the solar system.

Before Jewitt and Luu had breakfast that morning in Hale Pohaku, they called Brian Marsden at the Minor Planet Center at Harvard Observatory. He was of course very enthusiastic but emphasized that there was little that could be said with certainty about the exact orbit of this newly discovered object. At such a great distance from the Sun, its observed motion in the night sky was primarily a reflection of the Earth's orbital motion; it was therefore not clear if the object moved in a regular circular orbit or a very elongated ellipse. Perhaps Jewitt and Luu had discovered an extremely distant comet, or a third centaur.

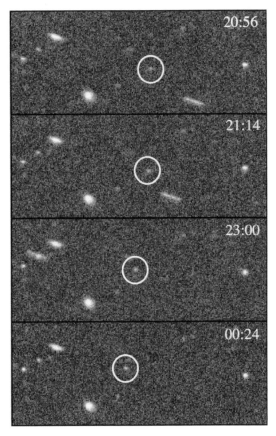

Discovery images of 1992 QB$_1$, the first Kuiper Belt Object (University of Hawaii/David Jewitt)

The remaining three nights of the observing session were devoted to new positional and brightness measurements for the new object. Dave and Jane called it Smiley, a humorous reference to George Smiley, the somewhat oafish secret agent in the John le Carré novel that Jane was reading at the time. But above all it was a cheerful name: it makes you smile– and they had reason to smile. A 7-year search had finally paid off. The outer regions of the solar system are *not* empty. The discovery of the Kuiper Belt seemed to have become a reality.

On September 14, Marsden announced the discovery of 1992 QB$_1$, the temporary designation of the new mini-planet, in circular 5611 of the International Astronomical Union. A week later, 'Smiley' was also observed by telescopes in Australia and Chile, and before long its orbit could be determined more precisely. The object proved to lie entirely beyond the orbit of Neptune. Its distance from the Sun varies from 6.1 to 7 billion kilometers (40.9–46.6 astronomical units [AUs]) and its orbital period is a little over 289 years. It was beyond doubt: Jewitt and Luu had found the first ice dwarf, the first *trans*-Neptunian body, and the first object in the Kuiper Belt. It was certain to be the first of many.

After the announcement of the discovery, the telephone in Honolulu never stopped ringing. There was considerable attention from the media and the usual questions were asked. Was this the long-sought Planet X? Could it support life? How many of these ice dwarfs were there? Would Smiley become the official name? And what went through the minds of the researchers when they realized what they had discovered? Luu and Jewitt had not expected so much interest; with hindsight it might have been better to have arranged an official press conference.

As far as the name was concerned it soon became clear that Smiley was not feasible. There was already a regular asteroid called Smiley (after astronomer Charles Smiley of Brown University) and according to the rules of the International Astronomical Union (IAU) committee responsible for naming small solar system bodies, each name could only be used once. Jewitt and Luu were, however, already not overly impressed by the IAU's bureaucracy and were not prepared to allow the Union tell them what to do, as a result of which 1992 QB_1 was given an official number (15760), but still has no name. 'Laziness has won,' admits Luu. Since the discoverer's right to name an object expires after 10 years, anyone can now propose an official name but the IAU committee has apparently not yet received any feasible suggestions.

In the meantime Jane Luu has almost completely turned her back on astronomy. After being awarded her PhD in 1992 she worked for a short time at the University of California at Berkeley and another couple of years at Harvard, but in 1997 she narrowly missed out on an attractive postdoc position at the Caltech. Caltech's Division of Geological and Planetary Sciences preferred Mike Brown, who later became famous as the discoverer of the tenth planet. After that, Luu spent 3 years at Leiden Observatory in the Netherlands (where she met her current husband Ronnie Hoogerwerf), but she returned to the United States in 2001, disillusioned. The weather in the Netherlands was worse than she ever could have imagined, European astronomy was shot through with nationalism and nepotism, and she found herself regularly at loggerheads with French colleagues about observing time on the European Very Large Telescope in Chile. She now conducts classified physical and technological military research for the MIT Lincoln Laboratory in Lexington. Most of her colleagues have no idea that she is the co-discoverer of the first ice dwarf, even if they know what the term means. But none of that matters much to Jane Luu.

Dave Jewitt cannot, however, get enough of the outer regions of the solar system. He has become an expert on the Kuiper Belt, has discovered a large number of new ice dwarfs, and has also been very successful in tracking down new satellites around the giant planets. But the discovery of 1992 QB_1 is the jewel in his crown – it is after all always nice to be the first to do something.

Or does the honor of finding the first ice dwarf belong to Clyde Tombaugh? After all, the discovery of Smiley and its fellow mini-planets made one thing clear: Pluto may be quite a lot larger than the other objects in the Kuiper Belt, but the 'ugly duckling' of the planet system is by no means unique. And if Ceres could no longer be called a planet after several dozen other asteroids had been found, then Pluto should suffer the same fate now that it is clear that thousands of other ice dwarfs are circling around in the outer regions of the solar system.

Dave Jewitt has never had the feeling that he has discovered Planet X. 1992 QB_1 is not the tenth planet, but the second ice dwarf. That is a shame for Billy and perhaps difficult for his father to swallow, but it is no longer acceptable to continue calling Pluto a planet. Is that a bad thing? The discoverer of Smiley cannot suppress a smile. No, it is not a bad thing at all. We may have lost a planet, but we've gained an entire Kuiper Belt in return. Surely that's a pretty good deal?

Chapter 18
Family Portraits

Chad Trujillo has been sitting for an hour, talking quietly about Pluto, the Kuiper Belt, and the hunt for Planet X. It is June 2006 and he has given up his free Saturday to come to his office. The corridors of the headquarters of the Gemini Observatory, in a suburb of Hilo, are dark and deserted. Chad's small room looks as though it has been hit by a tornado. Surrounded by untidy piles of books, magazines, and papers, he sits back in a comfortable chair and talks about the discovery of ice dwarfs with unusual dimensions and moving in strange orbits. He has a soft voice and friendly eyes, his hair is combed back in a ponytail, and his hands are folded calmly across his lap.

Then suddenly his small body starts to move, his dark eyes begin to sparkle, and there is passion in his voice. He is not talking about the discovery of 1996 TL_{66}, or a possible tenth planet, but the conversation has turned to children. Chad Trujillo became a father – for the first time – 3.5 months earlier. And the solar system simply cannot compete with his son Evan. Discovering a new planet is pretty cool, but it is nothing compared to the birth of your own child. No book or television documentary could have prepared him for the overwhelming experience of seeing his wife Tara bring a new human being into the world.

Chadwick Trujillo himself first saw the light of day at the end of 1973, somewhere near Chicago. He was just seven when Carl Sagan's series *Cosmos* was on the television, but the mysterious pictures of distant stars and planets made a profound impression on him. And during the twice-yearly outings with his parents to Pagosa Springs in Colorado, far away from the light of the big city, he could gaze breathlessly at the night sky. Add to that a knack for mathematics and it should come as no surprise that, while studying physics at the Massachusetts Institute of Technology, he specialized in astronomy.

Luck was with him in 1995 when he was admitted to the University of Hawaii in Honolulu to study a Masters in astronomy. What better place to gaze at the stars? The university was allocated 5–10% of the observing time of *each* telescope on Mauna Kea, including the enormous 10-meter Keck Telescopes, the second of which became operational shortly after Chad arrived. But using the university's own 2.2-meter telescope, at the highest point in the Pacific – 4,200 meters above sea level – was a wonderful experience.

G. Schilling, *The Hunt for Planet X*, DOI 10.1007/978-0-387-77805-1_18,
© Springer Science+Business Media, LLC 2009

Chad Trujillo with his son Evan (Courtesy Chad Trujillo)

And Dave Jewitt was a great teacher. When Trujillo took the plane to Hawaii, he knew nothing of the Kuiper Belt. The discovery of 1992 QB_1 in the summer of 1992 had completely passed him by. But under the palm trees of Honolulu – in sandals, shorts, and Hawaiian shirt – he was soon brought up to speed. Dave Jewitt and Jane Luu urgently needed software that would automatically look for slow-moving ice dwarfs in electronic images and Chad Trujillo seemed the perfect person to develop it.

Plutinos

Three years after the discovery of QB_1, as the first ice dwarf is usually called for short, study of the outer regions of the solar system had moved into a higher gear. Dozens of ice dwarfs similar to QB_1 had been found and it was clear that they were literally the tip of a gigantic iceberg. Beyond the orbit of Neptune there must be hundreds of thousands of these frozen mini-planets in orbit. To gain a better understanding of this new family, it was necessary to conduct the largest census possible, so that the statistics were reliable.

At first, progress was slow. Jewitt and Luu did not find the second ice dwarf – 1993 FW – until the end of March 1993 and, because they could not observe it long enough to calculate an accurate orbit, they lost sight of it again. The same happened with numbers three and four (1993 RO and 1993 RP), which they discovered in the first half of September. Because their orbits are not exactly known, these three objects have never been given an official number.

But Jewitt and Luu did not, of course, have sole rights to the Kuiper Belt. After the first fish had been caught, other researchers also cast out their lines.

The fifth and sixth ice dwarfs (1993 SB and 1993 SC) were found by Iwan Williams of Queen Mary University in London on September 16 and 17. Williams had always had a great interest in the origins of the solar system and was an expert on meteors and comets. It was therefore not a great step to explore the world of ice dwarfs, which are in fact 'supercomets' of a kind. Immediately after the discovery of QB_1, Williams and his colleagues Alan Fitzsimmons and Donal O'Ceallaigh requested observing time on the 2.5-meter Isaac Newton Telescope on La Palma, in the Canary Islands.

This time the orbits of 1993 SB and 1993 SC could be determined precisely (the two ice dwarfs were given the official numbers 15788 and 15789) and it was immediately obvious that something strange was going on. Both objects were at approximately the same distance from the Sun as Pluto (a little less than 6 billion kilometers) and as a result, they also had practically the same orbital periods as the outermost planet: between 240 and 250 years. That does not mean that their orbits are the same as that of Pluto: 1993 SB's orbit is much more elongated, while that of 1993 SC is more circular. Moreover, the orbits lie in different planes and have a different orientation in the solar system. But the similarities in average distance and orbital period immediately attracted attention: the two ice dwarfs discovered by Williams and his colleagues moved in Pluto-like orbits.

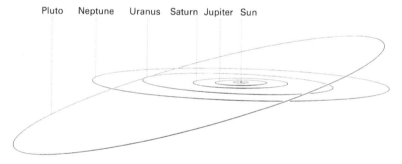

Pluto   Neptune   Uranus   Saturn   Jupiter   Sun

Pluto's orbit is very elongated and inclined with respect to the orbits of the other planets (Wil Tirion)

It was already clear shortly after Pluto was discovered that there was something extraordinary about its orbit. It is not a neat circle – its distance from the Sun varies from 4.4 to 7.4 billion kilometers – and it is at a pretty steep angle to the orbits of the other planets. But even more strangely, Pluto's orbital period is one and a half times longer than that of Neptune: in the time it takes for Neptune to go around the Sun three times, Pluto completes two orbits. This orbital resonance is of course not coincidental. The two planets feel each other's gravitational pull and because Pluto is periodically attracted to Neptune in the same way, their orbits remain 'linked.'

There are instances of orbital resonance elsewhere in the solar system. The orbital period of Jupiter's moon Europa is exactly twice as long as that of the moon Io, which is closer to the giant planet. The orbital periods of Saturn's moons Titan and Hyperion are in a ratio of three to four. And in the asteroid belt between the orbits of Mars and Jupiter, a relatively large number of these flying boulders have an orbital period which is two-thirds that of Jupiter. Computer simulations show that such situations are exceptionally stable: once two celestial bodies achieve orbital resonance, it can persist for a very long period. Evidently, the 3:2 orbital resonance with Neptune not only applied to Pluto – 1993 SB and 1993 SC displayed exactly the same 'commensurability,' as it is known in the jargon of celestial mechanics.

We now know that the same applies to 1993 RO and 1993 RP, the two ice dwarfs discovered a little earlier by Jewitt and Luu. If you come across four Pluto-like ice dwarfs so early in your search, Jewitt thought, there must be a very large number of them. He realized that there must be a separate family of ice dwarfs, of which Pluto was by far the largest, and called them plutinos – little Plutos. For Jewitt this made it even more crystal clear that Pluto is nothing more than an overgrown ice dwarf and that, for 65 years, the distant world had been erroneously classified as a planet.

Ice dwarfs have now also been discovered with a 2:1 orbital resonance with Neptune, giving them an orbital period of around 330 years. Other orbital resonances (4:3, 5:3, etc.) also occur, but the name 'plutino' is reserved for bodies with a resonance of 3:2.

And how about 1992 QB$_1$ and 1993 FW? They are on average further away and therefore have a longer orbital period of about 290 years. They have no special orbital resonance with Neptune. These 'classical' Kuiper Belt objects are more numerous than the plutinos and display a wide variety of orbital periods. Most of them are an average of 6.3–7.2 billion kilometers from the Sun (the actual distance can be considerably larger or smaller depending on the eccentricity of their orbits). Average distances of more than 7.5 billion kilometers (which correspond to orbital periods of about 350 years) are, remarkably enough, very rare. The Kuiper Belt therefore seems to have quite a sharp 'edge' at around 50 astronomical units (AUs) from the Sun. Since 1992 QB$_1$ was the first member of this large family to be discovered Brian Marsden suggested calling the classical Kuiper Belt objects *cubewanos*.

Speed Blinking

In short, Chad Trujillo's knowledge of the Kuiper Belt needed quite a lot of brushing up. Ice dwarfs, plutinos, orbital resonances, *cubewanos* – and the search continued unabated. Jewitt and Luu had had enough of 'blinking' their digital images. It was high time to fully automate the process.

It was of course not the first time that software had to be written to track down moving celestial bodies. The asteroid community had been using what was known as 'moving object detection software' for some time, but Trujillo decided to start with a clean slate. Ice dwarfs are not asteroids and they behave completely differently. During the exposure of an image, they hardly move, meaning that they remain point-like on the photograph. Asteroids on the other hand move so fast that they sometimes show up as elongated stripes. The difference in position of ice dwarfs between different images is also relatively small, requiring a different search strategy.

It was an interesting project, which would ultimately form the core of Chad's PhD thesis. The software was tested, improved, tested again, expanded, made smarter and faster, tried out on different telescopes and, in the meantime, used constantly in practice: first as a kind of independent check of the blink results, but later as the primary method for identifying new ice dwarfs (although the results were always checked visually). It then became a matter of taking the photographs, running the software and, before you knew it, you had another handful of mini-planets. This automated search method soon came to be known as speed-blinking.

But of course it was not all as easy as it sounds. Jewitt, Luu, Trujillo, and Jun Chen (a student born in China who had joined the team a little earlier) had no access to big computers on Mauna Kea. And naturally you want to unearth a new ice dwarf as soon as possible so that you can do follow-up observations in the same session. At that time laptop computers were not yet powerful enough to deal with the few gigabytes of measurement data gathered every night. So 15 PCs with monitors and extra hard disks had to be packed up in Honolulu for each observing session and taken along to Big Island, where they were set up in Hale Pohaku. It was no simple task.

The delight was therefore even more understandable when, on October 9, 1996, thanks to Chad Trujillo's brand-new moving object detection software, a very special object was found that appeared not to fit into any existing category. At least, that was what emerged in the weeks and months after it was discovered. At first, 1996 TL$_{66}$ appeared to be just another ice dwarf – a frozen mini-planet a few hundred kilometers in diameter at a distance of a little over 5 billion kilometers from the Sun. But when new positional measurements were conducted and Carl Hergenrother of the Lunar and Planetary Laboratory performed follow-up observations in December using a 1.2-meter telescope in Arizona, it soon became clear that the object had a very elongated orbit, with its furthest point at almost 20 billion kilometers from the Sun (131 AU)!

1996 TL$_{66}$ was certainly not a plutino and, with its enormous orbit and corresponding orbital period you could hardly call it a classical Kuiper Belt object. To acquire access to more positional measurements, Brian Marsden called in the help of 78-year-old amateur astronomer Warren Offutt from Cloudcroft, New Mexico, who had previously photographed ice dwarfs with a CCD camera mounted on his 60-centimeter telescope. In the night of January 10, 1997 Offutt succeeded in photographing the faint point of light, allowing Marsden to calculate its definitive orbit.

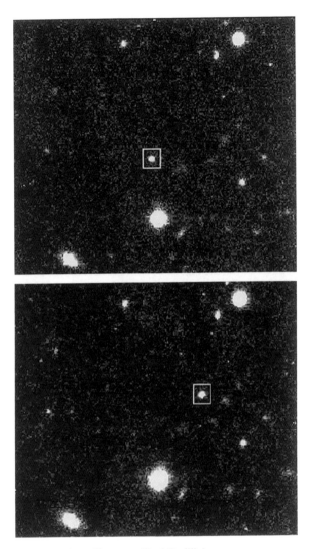

Discovery images of 1996 TL$_{66}$ (Courtesy Chad Trujillo)

No one had ever seen an orbit like this. TL$_{66}$ moves in an elongated ellipse at an angle of 24 degrees to the orbital plane of most planets. At the time of its discovery the ice dwarf was in the part of its orbit – at 5.25 billion kilometers distance (35 AU) – that was closest to the Sun: a little beyond the orbit of Neptune. But in the early seventeenth century, shortly after the discovery of the telescope, it was at the furthest point of its orbit – 19.5 billion kilometers from the Sun, more than three times the average distance of Pluto! With its orbital period of 755 years TL$_{66}$ was by far the slowest object in the solar system, with the exception of a few long-period comets.

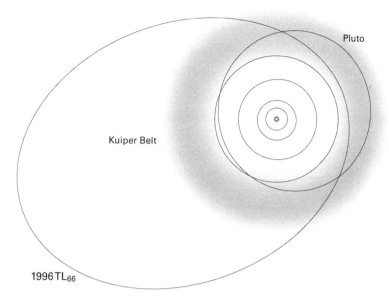

1996 TL$_{66}$ orbits in an extremely wide, elongated orbit around the Sun, well beyond the Kuiper Belt (Wil Tirion)

Mini-Feud

In the second half of January 1997, Luu, Marsden, Jewitt, Trujillo, Hergenrother, Chen, and Offutt submitted an article on 1996 TL$_{66}$ to the editors of *Nature* and at the end of the month Marsden announced the discovery through the *Minor Planet Electronic Circulars*. The *Nature* article was accepted for publication in mid-April and appeared in the June 5th issue.

That same spring a mini-feud broke out in the Kuiper Belt community. Eight days after the *Nature* article by Jane Luu and her colleagues was published, an article by Martin Duncan and Hal Levison appeared in *Science* predicting and explaining the existence of ice dwarfs with such large, elongated orbits. It was as though they wanted to let their colleagues in Hawaii know that the discovery of 1996 TL$_{66}$ was simply observational confirmation of their theoretical prediction. Luu was furious. It looked as though the theoreticians were going to get all the credit, while in her view it was the observers who had completely unexpectedly stumbled on a new type of celestial body. Furthermore, *Science* quite properly reported that the article by Duncan and Levison had been received on February 3, 1997, 4 days after Marsden's circular. The conclusion seemed obvious – the authors were playing dirty.

Martin Duncan can laugh about it now, although he and Luu have never buried the hatchet. Of course he understands that it is nicer to discover something that no one has thought of before but, in his view, that wasn't how it happened. Hal and he had been preparing their *Science* article long before Marsden announced the discovery of 1996 TL$_{66}$ on January 30. And well before the discovery, Jane Luu herself had taken part in an animated discussion on Duncan and Levison's theoretical ideas during a Kuiper Belt meeting at the Canadian Institute for Theoretical Astrophysics.

As with Duncan's work on the origins of the Oort Cloud and of the short-period comets, those ideas were based on extensive computer simulations of the evolution of the solar system. Simulations which showed that there must be a *third* family of ice dwarfs, in addition to the 'classical' Kuiper Belt objects (Marsden's *cubewanos*) and the resonant objects, of which the plutinos were by far the most numerous. That third family consisted of outcasts and pariahs: ice dwarfs which had been expelled from the actual Kuiper Belt in a cosmic diaspora by gravitational disturbances from Neptune.

If Neptune's gravity wreaked havoc in the Kuiper Belt, there were basically three possibilities, according to Duncan and Levison. An ice dwarf can be forced into a smaller orbit and – perhaps after an encounter with Uranus – end up as a centaur, like the extraordinary object Chiron discovered in 1977 by Charles Kowal. Or it can be expelled into a very elongated orbit where it will fall victim to the influence of passing stars and the tidal effects of the Milky Way. The third possibility is that the smaller bodies are catapulted out of the solar system but not far enough to actually leave it completely. Instead, they remain in an elongated and acutely angled orbit, with its nearest point close to the orbit of Neptune.

1996 TL$_{66}$ is a diaspora ice dwarf that is probably a member of an enormous family of *scattered disk* objects – bodies that have been flung out in all directions from the flattened disk of the Kuiper Belt by gravitational disturbances, but which are still a part of the solar system. It must be an extraordinarily large family, because if TL$_{66}$ had been in the outer part of its orbit (where it spends by far most of its time), it would have been at such a great distance from the Sun that it would never have been visible with ground-based telescopes. The fact that a diaspora ice dwarf could be found so shortly after the start of the search – because it happened to be at the closest part of its orbit – can only mean one thing: the 'scattered disk' must contain hundreds of thousands of similar objects, which together contain more ice than is currently present in the classical Kuiper Belt. Who knows what large objects might be hiding out there?

Chad Trujillo has never lost sight of the ice dwarfs since. Before he had been awarded his Masters degree, he had made his first major discovery and published his first article in *Nature*. A few years later he would join forces with Mike Brown and David Rabinowitz to search for the biggest ice dwarfs in the Kuiper Belt – Pluto-killers and Planet X's. And with success – but that is for a later chapter.

Chad Trujillo at a Kuiper Belt conference in Sicily, in 2006

These days, Trujillo primarily studies the composition of the distant mini-planets using the 8.2-meter Gemini North Telescope. Like their extraordinary orbital characteristics, the spectroscopic measurements of these bodies provide information on the origin and evolution of Kuiper Belt objects and therefore on the origins of the solar system itself.

Family portraits. Plutinos, *cubewanos*, centaurs, scattered-disk objects – anyone who studies facial expressions, movement, and body structure will eventually trace family trees and backgrounds. With little Evan it is easy. Chad saw him come into the world and he is clearly his and Tara's child. With Pluto, QB_1, Chiron, and TL_{66} it's a little more complicated. That calls for scientific detective work. A little less overwhelming and impressive than the birth of a child perhaps, but certainly just as fascinating.

Chapter 19
The Migrating Planet

At the end of January 1986, Miranda was the undisputed star of the Voyager show. Scarred landscapes, ice cliffs 20 kilometers high, a patchwork of the most diverse kinds of terrain – the small Uranian moon, discovered in 1948 by Gerard Kuiper, looked like a geologically deformed celestial body, a monstrous failed experiment in plastic surgery in the solar system. How did Mother Nature manage to create such a freak? It was as though Miranda had once been smashed to pieces and all the fragments stuck back together haphazardly. Planetary scientists at NASA's Jet Propulsion Laboratory in Pasadena were lost for words when the bizarre images from Voyager came up on their computer screens one after the other.

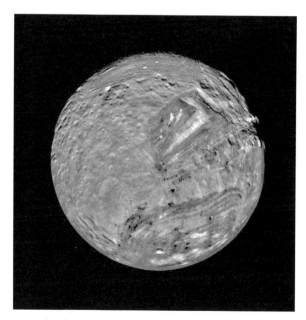

Voyager 2 photo mosaic of the Uranian moon Miranda (NASA/JPL)

G. Schilling, *The Hunt for Planet X*, DOI 10.1007/978-0-387-77805-1_19,
© Springer Science+Business Media, LLC 2009

Voyager 2, launched on August 20, 1977, had already flown past the giant planet Jupiter on July 9, 1979 and then Saturn on August 27, 1981. These flybys not only produced magnificent close-ups but also gave the probe extra speed. Thanks to the slingshot effect of their gravity, Voyager was able to travel the enormous distance to the next giant planet, Uranus, in only 4.5 years. On January 24, 1986 it passed a little more than 80,000 kilometers above the planet's cloud layers, sending back images of its atmosphere, moons, and rings. The pictures were spectacular, but Miranda was the jewel in the crown.

Planetary satellites are sometimes much more diverse than the planets themselves. Io, one of Jupiter's moons, has dozens of active sulfur volcanoes. Europa, a little further away from the giant planet, is an ice world with cracks and fissures and a deep ocean beneath its surface. On Saturn's large moon Titan it rains liquid methane gas. Enceladus has water geysers with which it sprays its neighbors in the cosmos, and Hyperion is a tumbling world of ice and rock that is as porous as a sponge. And to cap it all, Triton, Neptune's large moon, rotates around the planet in the wrong direction.

When it comes to explaining all these wonderful characteristics, geologists and astronomers come up with all kinds of weird and wonderful theories. Any theorist with a little creativity can find a solution for anything. But the more puzzling the observed reality, the more far-fetched the explanation. It has been suggested, for example, that Pluto was once a Neptunian moon, but was catapulted out of its orbit by the gravitational pull of a rogue planet. The same intruder allegedly forced Triton into its retrograde orbit. All possible on paper but very unlikely in practice.

Could there perhaps be a slightly less exotic explanation for the amazing appearance of Miranda (a name which very aptly means 'worthy of admiration')? That was the question that occupied a young Indian physics PhD graduate in the summer of 1986. On the advice of her supervisor Stanley Dermott, Renu Malhotra of Cornell University in Ithaca studied orbital resonances in the Uranian satellite system. They may have been indirectly responsible for Miranda's battered appearance. Io, for example, has to endure strongly varying tidal forces from its mother planet, as a consequence of similar resonances in Jupiter's satellite system. As a result of these tidal effects, Io's interior has melted and the relatively small body displays the most intense volcanic activity of any object in the solar system. A similar process could perhaps have caused drastic geological activity on and inside Miranda.

Odyssey

Renu Malhotra had dreamed of moving to the United States since she was 14. The land of progress, freedom, and unbridled opportunity! She could see it happening around her: if you had a good set of brains and a healthy dose of perseverance, American fortune would smile on you. The brains were no

problem. When Renu's poverty-stricken parents moved from their small country village to Hyderabad at the end of the 1960s, her mother succeeded in getting
her promisingly bright daughter admitted to the local convent school without
having to pay the tuition fee. An inspiring mathematics teacher did the rest.

Renu Malhotra as a high school student in India (Courtesy Renu Malhotra)

At the Indian Institute of Technology Delhi Renu was one of the five or six
female physics students among around 250 bright young Indian men. But she
had the full support of her parents and teachers, and when she was awarded her
Bachelor's degree in 1983, she felt the tug of the American dream. A few months
later she found herself at Cornell University, wracking her brains on mathematical problems of non-linear dynamics and working together with later Nobel
Prize winner Kenneth Wilson and Mitchell Feigenbaum, one of the founders of
chaos theory.

Malhotra's odyssey did not end in Ithaca. After fulfilling a postdoc position
at the California Institute of Technology in Pasadena, she worked for 9 years as
a staff scientist at the Lunar and Planetary Institute in Houston, where she
applied her knowledge of migration and resonances to the puzzle of Pluto's
orbit – and gave birth to two daughters, Mira and Lila. And in 2000, Malhotra's
own migration led her to Tucson, where she is now Professor of Solar System
Dynamics at the Lunar and Planetary Laboratory of the University of Arizona.

If anything resonates here, it is the almost tangible memory of Gerard Kuiper, founder of the laboratory, discoverer of Miranda, and the man who gave his name to the Kuiper Belt and the Kuiper Space Sciences Building, which houses Renu Malhotra's office.

In 1951, when Kuiper committed his ideas on the origins of the solar system to paper, he did not know what to make of Pluto. Such a small planet in such an elongated, inclined orbit – could Pluto have been created in the same way as the other planets? Or was there perhaps a core of truth in the theory suggested by the British astronomer Raymond Lyttleton in 1936, that Pluto had once been a Neptunian moon? On the other hand, the *closest* planet to the Sun, Mercury, also has quite an eccentric orbit which is inclined at a considerable angle. Perhaps, thought Kuiper, Mercury and Pluto, on the inner and outer edges of the protoplanetary cloud, were simply less sensitive to the regulatory effect of the other planets.

Renu Malhotra (University of Arizona)

Renu Malhotra believed none of this. The remarkable orbital resonance between Neptune and Pluto just cried out for a satisfactory explanation. At the moment, Pluto's orbital period is not *exactly* one-and-a-half times that of Neptune. Neptune goes around the Sun once every 164.774 years, while Pluto takes 248.026 years, 1.505 times as long. But extensive calculations and computer simulations show that the relationship between the orbital periods of the two bodies oscillates slowly around an average value of exactly 1.5. And if Pluto's average velocity were only a few meters per second faster or slower, its orbit would not be stable in the long term and the little planet would sooner or later be catapulted off into space. According to Malhotra, such a delicate dynamic relationship between two celestial bodies does not arise by chance.

So how did it happen? The migrant from India had only answer to that question: migration. She suggested that Neptune and Pluto first moved around the Sun at a much closer distance; in more or less circular orbits completely independently of each other. But for some reason Neptune moved further away from the Sun and its orbital period steadily increased. After a period of time, it was exactly two-thirds of that of Pluto and the two planets became trapped in mutual orbital resonance. And they stayed that way while Neptune's migration continued. As the giant planet moved slowly but surely further outwards, it was as though it pushed Pluto along in front of it.

Malhotra calculated that, if this were the case, the inclination and eccentricity of Pluto's orbit would automatically increase. So while Pluto gradually moved further away from the Sun, its orbital period remained on average one-and-a-half times that of the migrating planet Neptune, but its orbit would become steeper and more elongated. Pluto is like a piece of driftwood carried along on the waves of the passing cargo ship Neptune in the interplanetary ocean, and thereby forced to move a little more irregularly.

Migration

That is all well and good, but why should Neptune move outwards? Shouldn't planets just keep on following the same orbits round the Sun in perpetuity? In an ideal world perhaps they do, but the new-born solar system was far from ideal. Although the larger planets had already formed, there were still countless lumps of debris swarming around which would never come together to make a planet. Sooner or later most of these clumps of ice would be affected by Neptune's gravitational pull. And in reaction to this, in the early period of the solar system, Neptune must have gradually migrated outwards.

That a planet's orbit can be changed slightly by an encounter with another celestial body is nothing new. If Neptune can affect the orbit of an ice dwarf, the reverse is also possible. Of course, the motion of a small, low-mass ice dwarf can be much more easily disturbed than that of a massive giant planet, but there is definitely a reaction, no matter how small it may be. For example, Voyager 2 picked up a little extra speed when it passed close behind Jupiter in 1979, but the planet also lost a minute quantity of kinetic energy and moved into a very slightly smaller orbit around the Sun. Nature's energy bookkeeping is always very accurate, down to a whole series of decimal points.

But hold on a minute: if Voyager had passed Jupiter on the Sun side, it would have *lost* speed and the planet would have moved into a slightly *wider* orbit. So how does that work with Neptune and the ice dwarfs? The latter do not all fly past the planet in the same way: some move into a smaller orbit (where they end up as centaurs or short-period comets), while others move into much larger ones and find themselves in the 'scattered disk' or the Oort Cloud. Shouldn't the gravitational effects of these small bodies mean that Neptune would, on average, stay in exactly the same position?

That is exactly what Julio Fernández and Wing-Huen Ip expected in the early 1980s. Fernández, the Uruguayan astronomer who had already made a link between the Kuiper Belt and the short-period comets in 1980 (see Chapter 16), met Ip – who comes from Taiwan – while he was at the Max Planck Institute for Solar System Research in Katlenburg-Lindau, Germany, for a couple of years. In 1982 they conducted new and more detailed computer simulations which took account, for the first time, of the transfer of orbital angular momentum (a measurement of the kinetic energy of a celestial body in orbit around the Sun) between small ice dwarfs and large planets.

To their great amazement, the simulations showed that all four giant planets were seriously knocked off course by their interaction with the original ice dwarf population. Neptune and Uranus were forced quite a way outwards, Saturn was also forced into a slightly larger orbit, but to a lesser extent, while Jupiter moved a little closer to the Sun. That seemed to go against all common sense: shouldn't there be a balance between encounters which increased the planets' orbital angular momentum and those in which they lost some of their momentum?

If a computer simulation produces an apparently nonsensical result, the first thing you think of is a programming error. It took Fernández and Ip quite some time to convince themselves that there really was nothing wrong with the computer code. And it took a great deal of thought for them to understand the physical principle underlying their incredible findings. It meant working in reverse: mostly you conduct computer simulations to confirm ideas and theories, but Fernández and Ip had to wrack their brains to explain the results produced by the simulations.

To track down that explanation, you have to go back in time a few billion years. The giant planets have already formed, but there are still countless clumps of ice drifting around in the outer regions of the solar system – many of them close to the orbit of Neptune. Most of these are scattered by the effects of Neptune's gravity. About half of them end up in smaller elliptical orbits, the most distant points of which remain close to that of Neptune. The other half find themselves in *larger* elliptical orbits, with their closest points near to Neptune. The first group reduce Neptune's orbital angular momentum, the other increase it. The net result is no change. So far, everything is as it should be.

For the clumps of ice, however, it makes a great difference whether they are catapulted further away from the Sun or end up in a smaller orbit. Those that are expelled outwards spend most of their time at a great distance from the Sun, but never again come across a large planet, except the occasional encounter with Neptune. The second group, however, find themselves in the sphere of influence of Uranus and later perhaps of Saturn and Jupiter. And that is not such good news: the enormous gravitational pull of the giant planet Jupiter can easily cause an ice dwarf to accelerate so rapidly that it is catapulted entirely out of the solar system.

And that means that, as time passes, Neptune encounters fewer and fewer objects that initially ended up in smaller orbits, for the simple reason that a large number of them have been 'cleaned up' by Jupiter. Eventually, the remaining encounters are with ice dwarfs that have a larger specific orbital angular momentum, part of which is transferred to Neptune each time they pass close to the giant planet. The balance is therefore disrupted and Neptune migrates slowly outwards. The same applies to Uranus and Saturn. Only Jupiter moves inwards, as a result of its expelling all those unfortunate clumps of ice, some of which end up in the distant Oort Cloud, while the rest disappear forever into interstellar space.

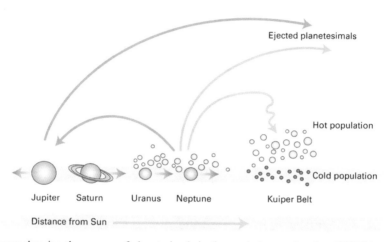

Diagram showing the process of planetesimal ejection and planet migration (Wil Tirion)

Resonances

An unexpected result and a surprising discovery: the clean-up operation in the early days of the solar system led to radical migration of the giant planets. Saturn, Uranus, and Neptune were formed closer to the Sun than their current positions, while Jupiter was originally a little further away. The migration did not come to an end until the planets had sufficiently cleaned up their own orbital regions. After the clean-up operation, the only ice dwarfs left were all *outside* the orbit of Neptune and the Kuiper Belt had largely acquired its present form.

You would expect planetary researchers to find the conclusions reached by Fernández and Ip extremely interesting. But the response was very lukewarm, despite the fact that their results had been published in 1984 in *Icarus*, the leading journal for solar system science. Just like Fernández' 1980 publication on the origins of short-period comets, almost no one noticed the *Icarus* article. It was in effect ignored for many years. Fernandez and Ip could simply have been ahead of their time: in 1984 no one was interested in the Kuiper Belt (the name had not even been thought of then), computer simulations were still in their infancy, and astronomers were perhaps not yet ready to consider the idea of giant planets losing their way in the solar system. Or perhaps the predominantly American community of solar system researchers were just not accustomed to the fact that Uruguayan and Taiwanese colleagues might also be able to make spectacular discoveries.

As an Indian, Renu Malhotra had no problems with nationalist sentiments, real or imagined. She knew the work of Fernández and Ip, respectfully referred to it in her own publications and refined their migration scenario with her own ideas on orbital resonance. During its migration, Neptune could also have swept up ice dwarfs in orbital resonances, which would increase their orbital inclination and eccentricity but would not affect the resonance between the ice dwarf and the giant planet. This was exactly what was needed to explain Pluto's strange and still unexplained orbit.

Malhotra published her theories on the origins of Pluto's orbit in *Nature* on October 28, 1993. And, unlike Fernández and Ip, she had no complaints about the attention and interest it received. The idea that the tiny planet's peculiar orbit had finally been explained more than 60 years after its discovery was very exciting. Furthermore, the outer regions of the solar system had become very fashionable: the series of articles by Martin Duncan, Scott Tremaine, and Thomas Quinn had put the Kuiper Belt on the map, the first *trans*-Neptunian objects had been discovered, and that had pushed Pluto back into the spotlight. In 1995, Malhotra published a long article in *The Astronomical Journal* and in 1997 she was awarded the Harold C. Urey Prize by the Division for Planetary Sciences of the American Astronomical Society.

Even better of course was that Malhotra's prediction that more Kuiper Belt objects would be discovered in orbital resonance with Neptune proved to be well-founded. Now it was known that the outer regions of the solar system were indeed populated by ice dwarfs, it was logical that during its migration Neptune would have swept up many more small bodies – not only in Pluto-style orbits with an orbital period one-and-a-half times that of the giant planet, but also in other resonances. In that respect, the discovery of the first plutinos at the end of 1993 was a fantastic confirmation of Malhotra's theory.

Since then, Kuiper Belt objects have also been found in a 1:1 resonance with Neptune – ice dwarfs with roughly the same orbital period of 164 years, even though their orbits may be more elongated and have a larger inclination than that of the giant planet. Like the plutinos and the other resonant ice dwarfs, they were probably forced into their strange orbits during Neptune's migration. So far, five of these 'Neptune trojans' had been discovered, but there are undoubtedly very many more – Jupiter, too, shares its orbit with countless small bodies, the largest of which have been named after heroes of the Trojan War.

By the middle of the 1990s it was in any case clear that the outer regions of the solar system had a tumultuous history, in which large and smaller celestial bodies performed a dynamic dance, orchestrated by gravitational disturbances, planetary migrations, and orbital resonances. Ice dwarfs were scattered throughout the universe, planets were knocked off course, and the current structure of the solar system – with its Oort Cloud, Kuiper Belt, scattered disk, centaurs, long and short-period comets, and mini-planets in resonant orbits – is the silent witness to far-reaching changes that took place in a distant past.

But anyone who wants to know more about the icy bodies beyond the orbit of Neptune must look further than their dynamical properties and their orbital evolution. At least as interesting is what they are made of, what materials exist on their surfaces, and how they come to have such a striking variety in color and brightness. Anyone who wants to really understand the Kuiper Belt, cannot do without a spectroscope.

Chapter 20
Icy Treasure Troves

They shoot into the universe in all directions. Left, right, up, down, north, and south. At the highest speed possible in nature: 300,000 kilometers a second. Photons from the Sun can cover the distance to the Earth in just over 8 minutes. Five and a half hours later they pass the orbit of Pluto on their way to the stars. They have been doing it for 5 billion years, and they will continue to do so for another 5 billion.

The Sun is a cosmic nuclear power plant with a capacity of almost half a quintillion gigawatts. That unimaginable energy radiates from the Sun's surface in the form of infrared rays, visible light, and harmful ultraviolet and X-rays. Countless photons (the light particles or energy packages first described more than a century ago by Albert Einstein) cross the almost-vacuum of interplanetary space at the speed of light in a wide range of all conceivable wavelengths.

To be completely exact, they do not exist at *all* wavelengths. Atoms in the outer gas layers of the Sun absorb a small part of the light. Each atom does that at its own specific wavelength. You don't actually notice it at all; it is only if you study the various wavelengths very closely that you will see that a few are missing. It is as though you remove three strings from a grand piano: Tchaikovsky's concertos sound just as impressive, but anyone who listens closely enough will discover that a few tones are missing.

The energy radiated by the Sun in 1 second would be sufficient to fulfill the needs of the entire global population for a million years. But of course, by no means all that energy reaches the Earth. The vast majority of photons make straight to the dark depths of the universe. From the Sun, the Earth – at a distance of 150 million kilometers – is an insignificantly small target. Our planet intercepts only one in every 650,000 photons emitted by the Sun – more than enough to prevent the oceans from freezing, unleash tropical hurricanes, and maintain all life on Earth.

For the distant dwarf planet Pluto, the situation is much less favorable. Pluto is far smaller than the Earth and much further away. No more than 1 in 30 billion photons from the Sun arrive at Pluto – not enough to thaw the ice world. The frozen planet absorbs about half of its incident light and the other half bounces back into space, causing certain wavelengths to be once again filtered

G. Schilling, *The Hunt for Planet X*, DOI 10.1007/978-0-387-77805-1_20,
© Springer Science+Business Media, LLC 2009

Artist impression of Pluto and Charon, as seen from the surface of one of the dwarf planet's smaller moons (NASA, ESA, and G. Bacon (STScI))

out by its surface material. That reflected sunlight is also dispersed through space and – after a journey of several hours – a minimal percentage arrives at our home planet.

It is not much, and the molecules in the Earth's atmosphere filter more of it out. The 'Pluto photons' then end up all over the Earth: in the oceans, the deserts, the tropical rainforests, and in the backyards of the readers of this book. Only a handful are captured by telescopes pointed precisely at the far-off mini-planet. Sensitive spectroscopes analyze the light, astronomers identify the missing wavelengths and, by taking account of the filter effect of the Sun and the Earth's atmosphere, they can determine the composition of Pluto's surface.

Spectroscopy enables astronomers to perform magic. In 1835 the French philosopher Auguste Comte stated that mankind would never be able to determine the composition of celestial bodies. A short time later, spectroscopy was invented, and enabled cosmic detectives to perform forensic analysis on starlight and therefore identify the fingerprints and DNA traces of atoms and molecules. At distance of hundreds of millions of kilometers or hundreds of millions of light-years.

But you need a lot of photons to perform spectroscopic analyses. A spectroscope splits the captured light up into all the colors of the rainbow; and if you want to measure how many photons come in for each wavelength, you need a good supply to work with. Furthermore, not all kinds of radiation are equally simple to split up. You can easily make a spectrum of visible light using a prism or a diffraction grating – a glass plate with very fine parallel lines engraved on it. (You can see how white light is split up into all the colors of the rainbow when the sun shines through a crystal chandelier. And anyone who doubts that very fine grooves can produce rainbow colors should try holding a CD up in the sunlight.) But X-rays and ultraviolet light are more difficult to break down, and spectroscopy is also a tough job on infrared wavelengths.

Pitfalls

Dale Cruikshank was very well aware of all these pitfalls. Pluto is a good 6 billion kilometers away from the Earth and a ground-based telescope can only capture a very small amount of light from the distant body. Furthermore, to determine the composition of its surface, you really need to analyze the infrared part of the light: in addition to frozen water, the surface may also contain methane snow or ammonia ice, and all these frozen substances leave their spectroscopic fingerprints most noticeably at infrared wavelengths.

But Cruikshank did not give up that easily. He had spent too much time in the vicinity of the legendary Gerard Kuiper, his intrepid hero and shining example. Dale was 14 when, in the early morning of June 30, 1954, he witnessed an impressive solar eclipse at his home in Iowa. Two years later, when Mars was close to the Earth, he read in the newspaper that Professor Kuiper of Yerkes Observatory in Wisconsin had observed clouds around the red planet. This shadow ballet of celestial bodies and clouds on a faraway, mysterious world fascinated Dale and he resolved to become an astronomer. In 1958 he applied for a summer job at Yerkes and a few weeks later the 18-year-old was in the dark room with Kuiper preparing images for a large photographic lunar atlas.

Cruikshank was just the kind of student Kuiper liked. He was in awe of his teacher, did exactly what he was told and remained polite and subservient under all circumstances. For his part, Kuiper was the driving force behind Cruikshank's career and, in fact, a determining factor in his life. In the early 1960s Dale moved to Tucson with Kuiper, where the Lunar and Planetary Laboratory was set up, and when he was awarded his PhD by the University of Arizona in 1968 he already had more than 20 scientific publications to his name.

Of course, Cruikshank's pioneering work in the infrared spectroscopy of Pluto was influenced by Kuiper. Kuiper had been working with infrared detectors and simple spectrometers since the end of the 1940s, with which he had made several discoveries, including the atmosphere around Saturn's moon Titan. It had never occurred to Cruikshank that he could follow in Kuiper's footsteps. The founder of American planetary science was far too big, too much in a class of his own. But as telescopes got larger and measuring instruments more sensitive, he saw more opportunities to make new contributions to the imposing edifice of Kuiper's research into the solar system.

The brainwave came in 1974, a few months after Kuiper's death. Cruikshank was now associated with the University of Hawaii. Infrared spectrometers were still not sensitive enough to analyze Pluto's weak light, but that was no reason not to try: you could also study Pluto's surface without a spectrometer. All you need is an infrared detector (a sort of light meter for heat radiation) and a set of six carefully selected filters. Seen through these filters, Pluto's infrared brightness tells you what kind of ice is present on its surface.

The idea is very simple. Laboratory measurements show that frozen water and methane ice absorb a lot of radiation at wavelengths between 3.2 and 3.7 millimeters. Ammonia crystals, however, absorb almost nothing at those wavelengths. If you therefore observe Pluto with a filter that only allows infrared radiation at those wavelengths to pass through it and the planet still remains clearly visible, there must be ammonia ice on its surface. Conversely, if the mini-planet is very dark, there is no ammonia ice, but water or methane ice is present. At other wavelengths, there is strong absorption from frozen ammonia and water, but not from methane. In this way, the search to determine the composition of Pluto becomes a kind of logic puzzle: if you combine enough different measurements, you can eventually deduce what materials combine to make up its surface.

Dale Cruikshank (left) and Carl Pilcher at the prime focus of the 4-meter Mayall Telescope at Kitt Peak National Observatory (Courtesy Dale Cruikshank)

Cruikshank joined forces with colleagues David Morrison and Carl Pilcher. The filters had to be specially made but, in March 1976, the observations could finally be carried out. They did not use the University of Hawaii's own 2.2-meter telescope on Mauna Kea, but the 4-meter Mayall Telescope at Kitt Peak National Observatory in Arizona. This telescope not only was quite a bit bigger, but also had better infrared detectors. Eight months later, on November

19, the astonishing results were published in *Science*: Pluto's surface is covered with a layer of methane snow.

As frozen methane is bright white, Pluto must reflect much more sunlight than had hitherto been assumed. To explain the level of brightness observed, therefore, the planet does not therefore have to be that large – Cruikshank and his colleagues estimated its diameter at around 3,000 kilometers. In the final paragraph of their article in *Science*, they therefore concluded that Pluto is too small and too lightweight to disturb the motions of Uranus and Neptune, and that Clyde Tombaugh's discovery in 1930 had more to do with his dedication and perseverance than with Percival Lowell's predictions about Planet X.

Pholus

Today, infrared spectrometers play a crucial role in planetary science. Pluto's reflection spectrum has been mapped in detail and the composition of its surface is now known much more accurately. Dale Cruikshank was part of a team that made precision measurements with an extremely sensitive infrared spectrometer on the 3.8-meter United Kingdom Infra Red Telescope (UKIRT) on Hawaii in 1993. The team, led by Tobias Owen of the University of Hawaii, discovered that frozen nitrogen occurs on the surface of Pluto in much larger quantities than methane ice. The surface of Pluto's large moon, Charon, on the other hand, proved to be made up primarily of frozen water and ammonia. The ice moons of the giant planets have since all been studied in detail and the infrared spectrum of many newly discovered objects in the Kuiper Belt has also been measured, especially if the objects attract particular attention for one reason or another.

David Rabinowitz, who discovered Pholus (Courtesy David Rabinowitz)

That was certainly true of Pholus, which was discovered on January 9, 1992 by David Rabinowitz of the University of Arizona. Rabinowitz had a postdoc appointment on the Spacewatch team initiated by Dutch–American

astronomer Tom Gehrels. At the end of the 1980s Rabinowitz developed a large
digital camera for the 90-centimeter Spacewatch Telescope on Kitt Peak. Using
the fully automated telescope, Gehrels scoured the heavens for unknown
comets and asteroids. The search produced countless Earth-grazers – small
asteroids which can pass close to the Earth and which, at some point in the
distant future, may collide with our own planet.

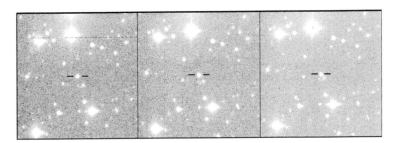

Discovery images of Pholus (University of Arizona)

In the case of Pholus (which was originally known as 1992 AD) we do not
need to be afraid of a collision. The object, with an estimated diameter of
185 kilometers, moves through the outer regions of the solar system in a very
elongated, strongly inclined orbit. The distance varies from 1.3 to 4.8 billion
kilometers (8.7–32 astronomical units [AUs]). Pholus is therefore classified as a
centaur: like Chiron (discovered by Charles Kowal in 1977) it almost certainly
comes from the Kuiper Belt but, as a result of the gravitational effects of the
giant planets, has now been forced into a smaller orbit, the closest point of
which is slightly inside Saturn's orbit. On the other hand, again like Chiron, its
orbit is not stable in the long term, so the possibility cannot be completely ruled
out that it might pose a threat to the Earth several million years in the future.

Perhaps equally as fascinating as Pholus' orbit was its striking red color,
which several astronomers noted immediately after its discovery. David Tholen
of the University of Hawaii even called Pholus the reddest 'asteroid' he had ever
seen. Tholen's colleague Gillian Wright recorded the centaur's infrared spectrum later that year with UKIRT and discovered tar-like organic compounds
on its surface which can, under the influence of ultraviolet sunlight, be formed
from molecules of methane and ammonia. Similar organic polymers have been
found on Saturn's moon Titan. In 1979, the famous planetary scientist and
popularizer Carl Sagan called them 'tholins' – not after David Tholen, but after
the Greek word for 'muddy.'

Even before the discovery of the first *trans*-Neptunian object (1992 QB$_1$) it
was therefore already clear that the cold exterior regions of the solar system
contained an enormous variety of bodies: Pluto, Charon, Chiron, and Pholus
all have their own characteristic surface properties with variations in reflectivity, color, and composition, which must in one way or another say something

about how these bodies originated and evolved. A red surface, for example, is generally assumed to point to a relatively advanced age: over a period of tens of millions of years, under the influence of ultraviolet sunlight and cosmic rays, new complex organic compounds (like tholins) can arise with a predominantly dark-red color.

Variety

The many hundreds of ice dwarfs now known display a comparable variety. That was clear as soon as the first infrared spectra of 'regular' Kuiper Belt objects were recorded. Three guesses as to who was the first person to record the infrared spectrum of one of these faint celestial bodies: Dale Cruikshank of course. Cruikshank now works at NASA's Ames Research Center in Moffett Field, California. At the end of 1996, together with his former student Bob Brown of the University of Arizona, Yvonne Pendleton from Ames, and Glenn Veeder from the Jet Propulsion Laboratory, he used the 10-meter Keck Telescope on Hawaii to observe 1993 SC with a sensitive infrared detector. The observations showed that there was probably a mixture of frozen hydrocarbons, including methane and perhaps also frozen nitrogen on the surface of this plutino. Apparently, not only did 1993 SC's orbit resemble that of Pluto, it also has a similar surface composition.

For the second and third ice dwarfs for which an infrared spectrum was acquired, however, that was by no means the case. Dave Jewitt and Jane Luu studied 1996 TL_{66}, the first object discovered in the 'scattered disk' whose orbit was predominantly far beyond Pluto. Like Chiron, 1996 TL_{66} proved to have a completely featureless spectrum, without conspicuous 'fingerprints' of, for example, water or methane ice. And 1996 TO_{66}, observed again by the team of Cruikshank and Brown, displayed clear absorption bands of frozen water and therefore more closely resembled the Plutonian moon Charon. Striking red-colored ice dwarfs have since been found, but Pholus remains the most spectacular. Cruikshank and his colleagues have also found frozen methyl alcohol (methanol) on this centaur – a molecule only found previously in comets. That supports the theory that there is an evolutionary link between ice dwarfs, centaurs, and short-period comets.

In less than a decade, the study of the surface composition of ice dwarfs has become a complete industry, for which the world's largest telescopes are used, such as the two 10-meter Keck Telescopes and the 8.1-meter Gemini North Telescope on Hawaii, and the European Very Large Telescope at Cerro Paranal in North Chile. The latter consists of four identical telescopes, each with a mirror 8.2 meters across. Nevertheless, we still know very little of the origins of all this variety. Planetary scientists of course do their utmost to discover links between the surface composition and orbital properties of ice dwarfs. Classical Kuiper Belt objects with low orbital inclination tend, for example, to be a little redder than those with a steeper orbit. But, as of 2007, there have been hardly any major breakthroughs or notable new insights.

And yet that minute amount of reflected sunlight, that handful of photons from billions of kilometers away, must contain evidence of the origins and early evolution of the solar system. Of course, you can deduce a lot from the orbits and the various families of ice dwarfs, but anyone wishing to obtain a complete picture of the inhabitants of the Kuiper Belt must find out more about their physical and chemical properties. After all, if you are studying the people of a neighboring country, you are not satisfied with only looking at where they live and their travel habits. You also want to know more – how tall they are, how much they weigh, and the color of their hair.

As long as astronomers have no readymade explanations for the great variety in the Kuiper Belt, they clearly lack an essential component in their ideas on the origins and evolution of the solar system. The number of newly discovered ice dwarfs beyond the orbit of Neptune had increased rapidly by the beginning of the twenty-first century, but that did not make things much clearer. And one burning question certainly remained unanswered: is there a Planet X floating around somewhere in the outer regions of the solar system – a tenth planet, larger than Pluto? And if so, who is going to find it?

Chapter 21
The Big Five

Mike Brown could not believe his ears. Was the renowned Schmidt Telescope at the Mount Palomar Observatory really going to be mothballed? Surely it was working just fine?

It was 1997, Brown had just been appointed associate professor at the California Institute of Technology (Caltech) and this was one of his first visits to the observatory in its idyllic location to the north of San Diego. That winter night, he was going to conduct observations with the 5-meter Hale Telescope. Unfortunately, it snowed nearly all night and visibility was zero. So he had plenty of time for sightseeing. And while Jean Mueller, of the observatory staff, was showing him the almost 50-year-old, 1.2-meter Samuel Oschin Schmidt Telescope, she happened to mention that it would probably be shutting down within a couple of years. It didn't bear thinking about!

Half a century is a respectable age for a telescope. But for a long time, the Palomar Schmidt – which was only named after businessman and sponsor Samuel Oschin in 1987 – was without equal. It is in fact an astronomical wide-field camera, which has been used to take many thousands of photographs of the night sky above South California. Countless asteroids, comets, and supernovas have been discovered with the telescope and Charles Kowal found Chiron with it in 1977. Were they going to pension off a telescope with such an illustrious track record? No way, thought Brown. He could use the Oschin Telescope to hunt for Planet X.

Little Michael knew from a very young age that he wanted to be an astronaut or an astronomer. He was brought up on space travel. His father worked on rockets for the Apollo program in Huntsville, Alabama. And when Mike had just turned four, he watched Neil Armstrong take the first steps on the moon on television at preschool. His high school was named after Gus Grissom, America's second astronaut, who died in the fatal fire in the Apollo 1 capsule in 1967. It therefore came as a surprise to no one when he went to study physics at Princeton, became interested in cosmology, and moved to the University of California in Berkeley in 1988 to do his Master's degree in astronomy.

G. Schilling, *The Hunt for Planet X*, DOI 10.1007/978-0-387-77805-1_21,
© Springer Science+Business Media, LLC 2009

The Samuel Oschin Schmidt Telescope at Palomar Observatory (Palomar Observatory/
California Institute of Technology)

Yet Mike Brown could not find his feet there. He was happiest trekking in
the mountains or sailing in San Francisco Bay. He didn't have a house; he lived
on his sail boat in Berkeley marina and once, when it had to go in for repairs, he
even slept for a month illegally between the small telescope domes on the roof of
the astronomy faculty. And he found it difficult to choose a subject for his
thesis. His supervisor Hyron Spinrad studied radio galaxies at cosmological
distances, but was also interested in comets. More or less by accident – and
completely against his own expectations – Brown found himself doing solar
system research, partly due to the enthusiasm of planetary scientist Imke de

Pater. In 1994 he was awarded a PhD for his study of electrically charged particles in the orbit of Jupiter's moon Io.

Mike Brown at a Kuiper Belt conference in Sicily, in 2006

But by then, Brown had already set his sights on the Kuiper Belt. During his PhD research he had a room next to Jane Luu, who was working in Berkeley as a brand-new postdoc and had just discovered 1992 QB$_1$ together with Dave Jewitt. Immediately before the discovery of the first *trans*-Neptunian object was announced, Luu showed Brown the images of the distant, moving point of light. It did not take him long to conclude that there must also be much larger bodies floating around in the outer regions of the solar system, perhaps even bigger than Pluto. Five years later, Mike Brown had his first encounter with the Samuel Oschin Telescope.

Large Field of View

As large ice dwarfs are much less numerous than small ones, you have to survey the largest possible part of the night sky. The Schmidt Telescope, with its photographic field of view of 6 by 6 degrees, was perfect for the hunt for the tenth planet. Brown did not mind that it could not detect the faintest Kuiper Belt objects. He was, after all, looking for the rare larger examples and they would of course be relatively brighter.

This was a completely different approach to the Deep Ecliptic Survey (DES), a search program that started in the fall of 1998 under the leadership of Bob Millis, Larry Wasserman, and Marc Buie of Lowell Observatory in Flagstaff.

DES uses a digital camera with 64 million pixels, mounted in the focal plane of the 4-meter Mayall Telescope at Kitt Peak National Observatory in Arizona or the practically identical Blanco Telescope at the Cerro Tololo Inter-American Observatory in Chile. Such a large telescope only has a small field of view: each DES image covers only a hundredth of a square degree. On the other hand, the combination of the telescope and the camera is extremely sensitive, allowing very faint ice dwarfs to be found. And because there are so many of them, the DES has found many hundreds, while Millis and his colleagues have covered only a very small section of the night sky. Good for the statistics, but not so useful if you are searching for rare objects.

That you need a large field of view to discover large ice dwarfs became apparent on November 28, 2000. On that night, using the University of Arizona's Spacewatch Telescope, Bob McMillan found Varuna, the first 'giant' among the ice dwarfs. The Spacewatch Project had been launched in May 1983, under the leadership of Tom Gehrels, who was primarily interested in tracking down bright Earth-grazers – asteroids which can pass dangerously close to our home planet. The first Spacewatch camera had less than 200,000 pixels (CCD technology was still in its infancy), but under McMillan's direction the 90-centimeter telescope was completely automated and equipped with a 4-megapixel camera with a field of view as large as the Full Moon. The camera was designed by David Rabinowitz, who had used it in early 1992 to discover Pholus, the unique centaur that spent most of its time between the orbits of Saturn and Neptune. It was inevitable that Spacewatch would sooner or later track down large ice dwarfs in the Kuiper Belt.

Varuna (which was originally called 2000 WR$_{106}$) is named after the most important god in the Hindu *Rig Veda*. It is about 900 kilometers in diameter, much bigger than the other ice dwarfs discovered until then. Surprisingly, DES found a *second* large ice dwarf 6 months later, on May 22, 2001, purely by coincidence. Ixion (2001 KX$_{76}$) is named after the king of the Lapiths in Greek mythology and probably has a diameter of around 800 kilometers. After the discovery of these two heavyweights, there could no longer be any doubt about it: the Kuiper Belt contains not only countless small ice dwarfs but also a large number of larger objects. For anyone capable of a little statistical extrapolation, it was crystal clear that one of these could well be a 'Pluto killer' – a Planet X.

It is not at all easy to determine the diameter of a small ice planet on the edge of the solar system. Even through the largest telescopes in the world, they appear as no more than a faint speck of light. Once the orbit has been determined, all you know is the distance from the Sun and the amount of sunlight reflected. You therefore have to make a guess about the reflectivity of the surface. The darker the surface, the larger the body must be to explain the observed brightness. In estimating the dimensions of Varuna and Ixion, planetary scientists assumed that, like Pluto, both objects reflected around half of their incident light.

These were exciting discoveries with fascinating names and they attracted a lot of media attention. But Mike Brown was of course not too happy about it.

He had been working with photographic plates on Palomar Mountain since 1998, just as Charles Kowal had done 20 years previously. The main difference was that the large Schmidt plates had now been digitized and software developed to search for moving objects. There was no longer any need for 'blinking' and everyone hoped that the computer would search more precisely than the easily distracted and exhausted human brain. But, although the software had often caused a false alarm as a result of dust particles or hairs on the plates, it had so far found nothing at all.

Quaoar

In the fall of 2000, Brown had given up the search. And he had neither the patience nor the energy to write a decent scientific paper about the negative results of his work. He much preferred to leave the past behind him and start afresh with a new search. And this time not with old-fashioned photographic plates which had to be exposed for an hour one at a time, but with sensitive electronic detectors, with which you could survey a large part of the night sky much more quickly. With the help of Chad Trujillo – who in August 2000, after being awarded his PhD, had gone to work as a postdoc at Caltech – Brown prepared a new survey, which would combine the large field of view of the Samuel Oschin Telescope with the sensitive 50-megapixel Near Earth Asteroid Tracking (NEAT) camera.

NEAT was a program operated by NASA's Jet Propulsion Laboratory in Pasadena (a few kilometers from the Caltech campus) to hunt for bright Earth-grazers. It initially used a 90-centimeter telescope owned by the US Air Force on the Haleakala volcano on Maui, Hawaii, together with a 16-megapixel camera that was revolutionary for its time. It had been developed by David Rabinowitz, the CCD wizard who had previously worked on Tom Gehrels and Bob McMillan's Spacewatch Project. But in 2000 Steven Pravdo's NEAT team had linked three cryogenic CCD cameras to create one massive detector with 50 million pixels. This 'three banger' was used successfully in combination with the Oschin Telescope on Palomar Mountain which, partly due to Brown's earlier photographic survey, had still not been taken out of operation.

Of course, Brown and Trujillo had to negotiate carefully with the NEAT team on use of the three banger, but that was not the only potential obstacle. The search for slow-moving ice dwarfs calls for a different strategy than the hunt for bright Earth-grazers flying past at high speed. It required completely new telescope control software and Trujillo also turned his attention to new moving-object-detection software. As a consequence, no observations were carried out at the end of 2000 and for most of 2001. This was exactly the time in which Varuna and Ixion stole the show as the first two ice dwarfs in the thousand-kilometer class. It was all very frustrating.

But once everything was up and running again in October 2001, success was not long in coming. On January 10, 2002 Brown and his colleagues discovered an ice dwarf with an estimated diameter of a little less than 800 kilometers, which was given the provisional name 2002 AW_{197} (and still has no official name); and on June 5 – Brown's 37th birthday – 2002 LM_{60} was found on images that had been recorded the day before. A better birthday present was hardly conceivable: with a diameter of 1,250 kilometers, the new object broke all existing records. It was half the size of Pluto and more massive than all the other known Kuiper Belt objects put together! No wonder that, after the discovery was announced on October 7, stories of the tenth planet started to crop up everywhere again, despite the fact that Brown and Trujillo insisted on their own websites that no astronomer would even think of calling the giant ice dwarf a planet. On the contrary: the discovery made it even clearer that it was time for Pluto to be demoted.

Chad Trujillo announced the discovery of 2002 LM_{60} during a meeting of the Division for Planetary Sciences of the American Astronomical Society in Birmingham, Alabama. The orbit had by then been determined accurately and Brian Marsden of the Minor Planet Center had given the object a permanent number: 50,000. It was no coincidence of course – this neat round number had been reserved for a special object for some time, and this object was definitely special. (Varuna had been given the number 20,000 for the same reason a few years before.) And since objects with a permanent number can also be given an official name, Trujillo could immediately reveal in Birmingham that the small ice planet was to be called Quaoar.

Artist impression of Quaoar (NASA)

Quaoar is the creation deity of the Native American Tongva people, who used to live in the area around Los Angeles and Pasadena a few centuries ago. Chad had always had a great interest in the culture of the original inhabitants of the United States, such as the Native Americans on the mainland and the Polynesians of Hawaii. In contrast to Greek and Roman mythology, these are at least cultures which still play an important role in the lives of the peoples concerned. Through the Tongva website, Trujillo made contact with tribal elder and historian Mark Acuña and they soon agreed that Quaoar would be a splendid name for the new body. The special committee of the International Astronomical Union responsible for giving names to small solar system bodies approved the name unanimously.

Quaoar's orbit could be determined so quickly because the mini-planet had been observed back in 1982 by none other than Charles Kowal – with the same telescope that Brown and Trujillo were now using. Kowal had not noticed the extremely faint speck of light (otherwise the Kuiper Belt would have been discovered 10 years previously!). That old positional measurement, however, made it possible to calculate Quaoar's orbit very accurately. It is practically circular and at an average distance of 6.5 billion kilometers from the Sun (43.4 astronomical units [AUs]).

The diameter, too, could be determined quite accurately. Through Steven Beckwith, director of the Space Telescope Science Institute, observations could be made with the Hubble Space Telescope in the summer of 2002. Because Hubble is not affected by the disturbing influence of the Earth's atmosphere, it has much sharper 'eyesight' than most telescopes on Earth. Quaoar proved just large enough to be seen as a tiny sphere. In this way it was possible to determine its diameter at 1,250 kilometers.

That was quite a lot larger than Brown and Trujillo had expected in the first instance. If, like Pluto, Quaoar reflected half of its incident sunlight, the observed brightness would have corresponded to a much smaller diameter. But the surface is apparently reasonably dark and reflects no more than 10% of the incident light.

Infrared observations with the Spanish IRAM Telescope in Granada confirmed this conclusion. Quaoar radiates a relatively large amount of heat with a wavelength of about 1 millimeter – exactly what you would expect from a dark object that absorbs a lot of sunlight and is therefore warmer than a reflective ice ball.

QUEST

Around the time that Brown and Trujillo announced the discovery of Quaoar, the first observations were made on Palomar Mountain with an even larger camera. It had been developed and built by astronomers and technicians from Indiana University in Bloomington and Yale University in New Haven – including of course David Rabinowitz, who had exchanged California for the East coast in 1999. The QUEST camera (the name officially stands for QUasar Equatorial Survey Team) is a kind of chessboard of 112 CCD detectors with a total of 161 million pixels on a surface about the size of a regular sheet of paper. In the focal plane of the Oschin Telescope, QUEST has a rectangular field of view of over 10 square degrees – three times the size of the three banger. And although the camera is predominantly used to survey quasars (extremely bright galaxies on the edge of the observable universe), it was also available from the summer of 2003 for Brown and Trujillo's planet hunt. As a member of the camera team, Rabinowitz had his own share of observing time, and Brown and Trujillo asked him to join the search.

The QUEST camera with its rows of CCD detectors. In the background is electronics designer Mark Gebhard (David Rabinowitz)

Little changed as far as the search method was concerned. It still entailed recording images of a specific part of the night sky three times in succession, with intervals of about 1.5 hours in between. In this period, because of the movement of the Earth, a Kuiper Belt object appears to have shifted by an average of 2–4 arc-seconds (1 arc-second is the diameter of a grain of salt seen from a distance of hundred meters). Special software is used to search the many gigabytes of telescope data for faint points of light that show consistent movement across the three separate images.

What did change was the extent to which the entire search was automated. Opening the dome, aiming the telescope, tracking the apparent nightly rotation of the sky, recording the images, and sending the data digitally to the computers in Pasadena all now take place without the intervention of a single human being. Only if it starts to rain on Palomar Mountain, does the night assistant on the 5-meter Hale Telescope have to press an emergency button to close the dome of the Oschin Telescope. With a rain sensor even that could be easily automated.

During the installation of the new camera in 2003, the hunt for large ice dwarfs had to be suspended for a few months but on July 26, with everything up and running again, the team immediately got lucky. On those first images, the bright ice dwarf 2003 OP$_{32}$ was found later in the year. 2003 OP$_{32}$ had an estimated diameter of a little under 700 kilometers. When the new software became completely operational in September, the discovery of new large ice dwarfs became almost routine. During the night, the computers churned out all their data and the following morning Mike Brown would be the first to look at the 10–20 candidates the software had detected. In most cases it was a false alarm, but very regularly one of them would turn out to be a new mini-planet.

Then, early in the morning, a cry of joy would resound through Caltech's South Mudd Laboratory, which houses the Division of Geological and Planetary Sciences. The door to room 168 was normally open and once, after she had heard him emit a primal scream three times in 5 minutes, Brown's PhD student Emily Schaller left her room across the corridor to ask him carefully if everything was okay.

On February 19, 2004, after more than 30 bright ice dwarfs had been found, Brown, Trujillo, and Rabinowitz announced the discovery of 2004 DW, which had been recorded 2 days earlier with the QUEST camera. It was a giant with an estimated diameter of 1,600 kilometers – considerably larger than Ceres (the record-holder in the asteroid belt between Mars and Jupiter), much larger than Quaoar and even quite a lot bigger than Pluto's moon, Charon. In November 2004 its orbit was determined accurately (2004 DW proved to be a plutino with an orbital period one and a half times that of Neptune) and the mini-planet was given the name Orcus, after a Roman god of the underworld.

Sedna

But while everyone was still responding enthusiastically to the discovery of Orcus – the fourth giant in the Kuiper Belt after Varuna, Ixion, and Quaoar – Mike, Chad, and David were busy preparing the biggest surprise of the year: the announcement of the discovery of Sedna, the most enigmatic celestial body ever to have been found in the outer regions of the solar system. Sedna made up the Big Five, but was also a painful reminder that the early evolution of the planetary system still held countless unsolved mysteries.

Discovery images of Sedna (California Institute of Technology)

Sedna was photographed for the first time on November 14, 2003, but remained elusive for quite a while after that. As with each new discovery, follow-up observations were made so that the orbit of the faint will-o'-the-wisp could be determined with greater precision. But the nights were either cloudy or the air was too turbulent, and 2003 VB_{12}, as the new body was called provisionally, appeared and disappeared unpredictably. Good reason for the three planet hunters to give it the nickname 'the Flying Dutchman.'

But the Flying Dutchman was located on a far horizon. The speed with which an ice dwarf seems to move in the sky is a direct indication of its distance. Sedna moved much more slowly than other ice dwarfs, suggesting that it must be 11 billion kilometers from the Sun – almost twice as far as Pluto! Never before had an object been found at such a great distance in the solar system.

Fortunately, as time passed, more and more positional measurements became available, also from the large telescopes on Hawaii and in Chile. And Sedna was identified on old images from 2001. That made it possible, in time, to determine its orbit with some degree of certainty. And to everyone's amazement, rather than being at the most *distant* part of its orbit, Sedna proved to be at the point closest to the Sun! Its extremely elongated orbit has its furthest point no less than 146 billion kilometers away – 975 AUs and 24 times further than the average distance of Pluto from the Sun. This gave the mysterious object an orbital period of some 12,000 years. (A small ice dwarf with an even longer period of 22,500 years was discovered in 2008 on images of the Sloan Digital Sky Survey).

Artist impression of Sedna (NASA/JPL)

Mysterious is the word. How does an ice dwarf end up in such a bizarre orbit? Sedna's orbit lies far beyond the Kuiper Belt and it is therefore not affected by the gravity of Neptune or any of the other planets. But that also implies that such planetary disturbances can never have caused the current shape and size of Sedna's orbit. Could there after all be a sizeable planet – a real Planet X – way out there, some 150 billion miles from the Sun? Or did Sedna originate in the interior of the Oort Cloud and its orbit was disrupted shortly after the birth of the solar system by nearby stars – perhaps the siblings of our Sun, all part of the same newly formed star cluster?

Whatever the solution to the mystery, this was clearly a revolutionary discovery. While the work with the QUEST camera continued, Brown, Trujillo, and Rabinowitz worked feverishly on a scientific paper for *The Astrophysical Journal* and consulted with Caltech's press office about the best publicity strategy. In the meantime, follow-up observations were planned with the IRAM Telescope in Spain and NASA's Spitzer Space Telescope. Neither succeeded, however, in detecting Sedna's radiant heat, which suggested that the distant mini-planet must be smaller than 1,800 kilometers across. On the other hand, Sedna appeared to have a striking red color, similar to that of the centaur Pholus, which led the team to suspect that its reflectivity was not extremely high. Sedna is therefore certain to be larger than Quaoar and possibly even slightly bigger than Orcus.

Sedna was the largest object to be found in the solar system since Tombaugh discovered Pluto in 1930. It was also the most distant object ever to be discovered in orbit around the Sun. And that orbit proved to have some very peculiar properties. The Caltech press release on the discovery was planned for Monday, March 15, 2004 (1 day before the article for *The Astrophysical Journal* was officially submitted) and, given the expected level of media attention, Brown and his colleagues decided to give the new object an interesting name. The provisional number 2003 VB_{12} was after all a bit of a tongue twister for radio and television reporters, and experience had showed time and time again that the general public loved names that appealed to the imagination.

Sedna is the sea goddess of the Inuit, the native inhabitants of Northern Canada and Greenland. As Sedna lives in the cold depths of the Arctic Ocean, it is very fitting that she give her name to a celestial body whose temperature never rises above −400 degrees Fahrenheit – only 60 degrees above absolute zero. Brown suggested the name to the International Astronomical Union and from March 15 started to use it for the intriguing object and had a separate Sedna page on his personal website, even though it had not yet officially been approved.

As far as media attention was concerned, Brown certainly had nothing to complain about. Sedna was in newspapers across the world, often on the front page. This fifth giant ice dwarf finally convinced practically everyone that, sooner or later, an object would be found in the outer regions of the solar system that was larger than Pluto. A tenth planet. And Brown was determined to be the one to find it. After all, he had the best camera, the most experience, and an exceptionally dedicated team. It would be frustrating if he were beaten to the post by other Planet X hunters. The clock was ticking.

What Mike Brown did not know at that moment was that he had already captured his much-desired prey on October 21, 2003. Planet X, the tenth planet, K31021C, the Pluto-killer, 2003 UB_{313}, Lila, Xena, minor planet 136199, Eris – or whatever else this elusive body has been called at different times – was already hidden on the hard disk of his computer. But it would not remain hidden for very long.

Chapter 22
The Tenth Planet

'I've just discovered a new planet.'

It was 11.30 in the morning on Wednesday, January 5, 2005. Mike Brown was pumped full of adrenaline. He could have run out of his office at the Caltech campus to shout the news from the rooftops. Or he could have sent an email to the science editor of the *Los Angeles Times*. But instead, he called his wife. 'That's nice, honey,' replied Diane Binney, who was 3 months pregnant with their first child. 'You won't forget to pick up some milk on your way home this afternoon now, will you?'

Artist impression of the 'new planet' discovered by Mike Brown (Robert Hurt/IPAC)

A small speck of light and a hardly observable shift in position, but nevertheless, a world-shattering discovery. That slow movement represented an

G. Schilling, *The Hunt for Planet X*, DOI 10.1007/978-0-387-77805-1_22, © Springer Science+Business Media, LLC 2009

enormous distance. Only a very large object could be so far away and yet still be so clearly visible. At a first guess, it was at a distance of about 18 billion kilometers, with a diameter of 6,000 kilometers – larger than the planet Mercury. There could be no doubt about it: this was the long-awaited Planet X. Michael E. Brown, 39 years old, had just found the tenth planet in the solar system, 75 years after the discovery of Pluto. He was writing history. Brown didn't give the milk one more thought that whole afternoon.

It had in any case been a sensible decision, in the spring of 2004, to write new software and go over all the old images recorded with the QUEST camera one more time. The most important reason for that was the discovery of Sedna: that bizarre object nearly twice as far from the Sun as Pluto and which therefore moves among the stars very slowly. If it had been a little further away, Sedna would not even have been detected by the old software, which only looked for objects moving at more than 1.5 arc-seconds an hour. The lower limit had been set to ensure that there were not too many false alarms, but it no longer seemed such a good idea now it was clear that there were large ice dwarfs far beyond Pluto's orbit. If Planet X were more than 12 billion kilometers from the Sun it could quite easily avoid detection.

From September 2004, therefore, the Caltech computers were extra busy. All the old QUEST images were analyzed again and, in the meantime, new data continued to come flowing in. And just beyond the orbit of Neptune, in the inner regions of the Kuiper Belt, the new software identified many more new ice dwarfs. Had the old version been written a little too hastily, as a result of which not all suspect points of light had been detected? Whatever the reason, it had missed quite a lot of moving objects.

That one of these objects was a relatively bright ice dwarf which also moved quickly through the sky was actually quite embarrassing. It was about 15 times as bright as Sedna and appeared to have a speed of about 3 arc-seconds per hour. An object like that should really have been found by the old software, though it was still 20,000 times too faint to be seen with the naked eye. It was located in the inconspicuous constellation of Coma Berenices, to the north of Virgo. It was the first Kuiper Belt object to be discovered on the QUEST images from May 6, 2004 and was therefore given the internal code K40506A, a very provisional name that was used only by Brown and his colleagues. But because K40506A was not actually discovered until December 28, just after Christmas, Brown gave it the nickname Santa.

It was immediately clear that Santa was a heavyweight. It was 7.8 billion kilometers from the Sun (52 astronomical units (AUs)), but was one of the brightest ice dwarfs ever discovered. Although the reflectivity of its surface was not yet known, it almost certainly had to be in the same category as Sedna. Brown emailed David Rabinowitz and Chad Trujillo to tell them about the discovery. In no time they dug up old images on which Santa was visible and booked observation time on large telescopes to find out more about this latest in a long line of hunting trophies. In the first few months of 2005, they found out that K40506A is peculiarly egg-shaped, rotates rapidly on its axis, and has a small moon.

Xena

Eight days after the discovery of Santa, while it was still relatively quiet at Caltech because of the Christmas holidays, Brown struck lucky again. This time he found something new on images of the constellation of Cetus dating from October 21, 2003. Thanks to the new software, he not only identified two new, faint ice dwarfs (K31021A and K31021B), but also a much brighter object that was given the internal code K31021C. It soon became clear why K31021C had escaped attention more than a year earlier: it moves very slowly (1.4 arc-seconds per hour), under the old software's limit of 1.5 arc-seconds per hour.

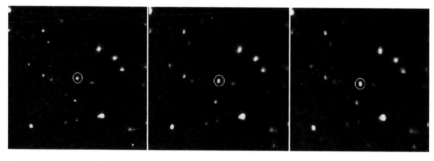

Discovery images of K31021C, alias Xena (Mike Brown, Chad Trujillo, David Rabinowitz/ Palomar Observatory)

If this new object moved so slowly it must also be at an inconceivably great distance from the Earth. Brown's original estimate of 18 billion kilometers proved a little high, but it was soon apparent that this was the most distant object in the solar system. K31021C was 14.5 billion kilometers from the Sun (97 AU) – almost three times as far as Pluto and 30% further than Sedna. Despite this, it was six times *brighter* than Sedna, which meant that it had to be much larger. Bigger than all the other ice dwarfs. Even bigger than Pluto. It was high time to call Diane.

Of course K31021C was also given a catchy nickname. Mike Brown didn't have to think about it too much – the name had been dreamed up a few years previously. He had spoken to Chad and David regularly about the possible discovery of a body larger than Pluto. The team had resolutely decided to call this potential first prize for planet hunters Xena, where the X was of course an unmistakable reference to Planet X, the unknown tenth planet. Xena was the pugnacious, voluptuous, and scantily clad main character from the cult television series *Xena: Warrior Princess*. The three astronomers still had pleasant memories of the series, which was broadcast between 1995 and 2001. After its discovery in June 2002, Quaoar had endearingly been known as 'mini Xena,' but now they had found the real Xena.

This was therefore great news, not only for Brown's wife, but especially for his team-mates Trujillo and Rabinowitz, who were once again informed by email. Rabinowitz immediately dived into the digital archives in New Haven and found Xena on QUEST images from a completely different observing program. He also excluded the unlikely possibility that Xena was actually a small asteroid in the inner solar system which was moving more or less towards the Earth and therefore appeared to move very slowly across the sky. Trujillo, too, studied old data and soon came across Xena in images from the 1990s, 1980s, and even eventually the 1950s, when the Schmidt Telescope on Palomar Mountain had just come into operation. It seemed unbelievable that no one had noticed the slow-moving point of light before.

All this detective work was conducted with the greatest secrecy, of course. As with earlier remarkable discoveries, like Quaoar, Sedna, and Santa, Brown had impressed upon his colleagues the importance of not publicizing their finds. If the news leaked out, someone else who happened to have access to a large telescope could cut the scientific ground from under the feet of Brown's team by being the first to conduct follow-up observations. It seemed more sensible to find out more about the orbit and physical properties of the new body themselves and then go public with the complete story. For the time being, they should tell as few people as possible about Xena's existence.

Complete secrecy was of course impossible. Like Brown, Rabinowitz and Trujillo told their wives. Chad Trujillo and his wife Tara had a lot of visitors at their home in Hilo during the Christmas holidays and it was difficult for them to keep the discovery of a new planet to themselves. And of course Trujillo had to tell his big boss Matt Mountain who, as director of the Gemini Observatory, had to give his personal approval for unscheduled observing time on the Gemini North Telescope. As well as his wife, David Rabinowitz also told his QUEST colleagues at Yale University, and Mike Brown told not only his mother but also Tom Soifer, director of the Spitzer Science Center, as he needed to use NASA's Spitzer Space Telescope to try and detect Xena's feeble radiant heat. Shortly after that, they also informed Steven Beckwith and Fred Chaffee, who were responsible for the Hubble Space Telescope and the Keck Telescopes on Hawaii.

In the meantime they had learned a little more about Xena's orbit. It was very elongated with a high inclination angle – typical of objects in the 'scattered disk.' Strangely enough, at the time it was discovered, Xena was at its furthest possible distance from the Sun: 14.5 billion kilometers. With an orbital period of 557 years, the mini planet will not reach the nearest point of its eccentric orbit – a 'mere' 5.6 billion kilometers from the Sun (37.4 AU) – until the second half of the twenty-third century. Xena's orbit is at an angle of 44 degrees to that of the Earth and the other planets, and can therefore almost never be seen in one of the constellations of the Zodiac. The same applies to Santa, which has an orbital inclination of 28 degrees. That was the reason why Charles Kowal did not find these bright ice dwarfs at the end of the 1970s; he directed his search for Planet X exclusively at the Zodiacal constellations.

Hubble Space Telescope image of Xena (NASA, ESA, and M. Brown (California Institute of Technology))

Tenth Planet

Xena therefore has a strange orbit, even more highly inclined and more elongated than that of Pluto, which was always considered the odd man out in the solar system. So could you in all decency call Xena a planet? After all, the 'normal' planets go around the Sun in more or less circular orbits and all in virtually the same plane. Mike Brown had a bit of a problem with that. Of course Xena was not a normal planet; it was part of the large family of *trans*-Neptunian objects, to which Pluto also belongs. When they discovered Quaoar and Sedna, Brown and his colleagues had insisted that these overgrown ice dwarfs were not real planets and that even Pluto's status as a planet was up for grabs. But the views of astronomers are not always shared by the public at large. Most people were not in favor of demoting Pluto and it is impossible to explain to someone who thinks that Pluto is a planet why Xena should not deserve the same status.

It was after all immediately clear that Xena is larger than Pluto. Once you know the distance and brightness, you can calculate the diameter. Of course, you also need to know how much sunlight the object reflects, but even if Xena had a reflectivity of 100%, it would be as big as Pluto. In practice, that never actually happens. The surface of a celestial object reflects perhaps 60 or 70% of incident light at best. Xena would therefore have to be quite a bit larger than Pluto to account for its observed brightness.

Although follow-up observations had to be carried out on Santa and Xena – the two giants among the ice dwarfs – in absolute secrecy in the first few months of 2005, the search with the QUEST camera continued unabated. And on Thursday, March 31, a third interesting Kuiper Belt object was found. It was a little brighter than Santa, but in a similar orbit and about the same distance from the Sun. Even more interesting, K50331A – as Brown called the object in his logbooks – was located in the same constellation as Santa: Coma Berenices. Because this discovery was made a few days after Easter, the object was given the name Easter Bunny.

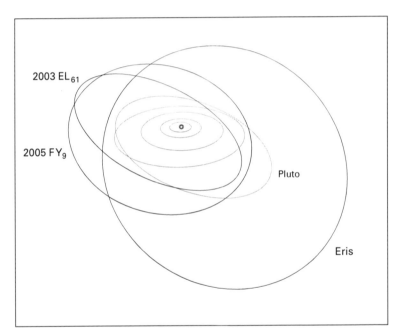

The orbits of Pluto, 'Xena' (now called Eris), 'Santa' (2003 EL$_{61}$), and 'Easter Bunny' (2005 FY$_9$) (Wil Tirion)

With the exception of Pluto, Easter Bunny, Santa, and Xena were the brightest ice dwarfs in the solar system. Together with Quaoar, Orcus, and Sedna they were the largest inhabitants of the Kuiper Belt. Xena was even slightly larger than Pluto and, with a little wishful thinking, you could call it the tenth planet. The discovery of these three record-holders was the crown of Brown's work. The search he had initiated 8 years previously with the Schmidt Telescope on Palomar Mountain had more than borne fruit. Thanks to large, sensitive CCD cameras, smart computer software, and an enormous dose of perseverance, Planet X had been found. The hunt was over.

But what should these new inhabitants of the solar system be called? Santa, Xena, and Easter Bunny were very clearly nicknames and far too frivolous to ever find their way into serious astronomical literature. K40506A, K31021C, and K50331A were code names used only by Brown, Trujillo, and Rabinowitz, for example, to apply for and process follow-up observations. Once the discoveries were reported to the Minor Planet Center, they would be given provisional designations – with a year, a letter, and a number code. But you couldn't impose these dry codes on the world. No, the new discoveries – especially Xena – needed proper astronomical names, preferably with a mythological tint.

Think up a name for Planet X. Why not? In Brown's household, there had been a lot of talk about names that spring. The first baby was on the way and Mike and Diane were busy drawing up lists of their personal favorites. There proved to be precious little overlap and their emotions could sometimes run high. Finally, the prospective parents decided on a compromise: if they couldn't find any names on each other's list that they really liked, they should at least mark all those that they didn't think were awful. So when Diane gave birth to a bouncing, healthy baby daughter in the afternoon of Thursday, July 7, she was given the name Lilah Binney Brown.

But now for the ice dwarf. New dad Mike, in his seventh heaven since the birth of his daughter, would have liked nothing better than to call the tenth planet Lilah, too. But it was hardly a mythological name. That had been a lot easier for Jim Christy in 1978 when he wanted to call Pluto's moon after his wife Char, and discovered that Charon was the name of the boatman of the underworld. Could Mike Brown come up with a similar convincing solution?

Lila

Where Christy had resorted to his encyclopedia 27 years earlier, in July 2005 Brown started with a small Google search on the Internet. And before long he found what he was looking for. In the Hindu faith, the universe is seen as a theater for the gods. This notion is described with a concept known as *Lila*, which is also the title of the second bestselling philosophical novel by Robert Pirsig, who had earlier achieved fame with his *Zen and the Art of Motorcycle Maintenance*. Admittedly, Lila is not a divinity, but it is not far off, and it did mean leaving the 'h' off the end of his daughter's name.... but all that was just detail. Mercury, Venus, Earth, Mars, Jupiter, Saturn, Uranus, Neptune, Pluto, and Lila – he liked the sound of that.

Brown realized that he would have to approach things a little more cautiously than he had over a year ago, when Sedna's discovery was announced. Then, the new object had been called Sedna from the beginning, even before the name had been approved by the International Astronomical Union (IAU) committee and the distant ice dwarf was still officially known only by its provisional designation, 2003 VB_{12}. That had caused irritation among other planetary scientists. The discovery had of course been big news in mid-March

2004 and received a lot of media attention. Everyone had used the proposed name and explained that Sedna was an important Inuit goddess. When the IAU committee came to give the object a permanent name, there was no longer any way around Sedna. In fact, critics claimed that Brown, Trujillo, and Rabinowitz had hijacked the name by announcing it prematurely. No one seemed to remember that Charles Kowal did exactly the same thing in 1977 with Chiron. There had been no protests at the time – Internet and email had not arrived and communication in the scientific world was slower and more amiable.

So in any case, it seemed better this time to adopt a little more caution, also because the name Lila would probably lead to more discussion in the committee. In other words, if the discovery of the tenth planet were to be announced, it would have to go on for a little longer under the name of Xena. No one could object to that, as long as it was clear that it was only a light-hearted nickname. The new object would be officially reported to the Minor Planet Center and the name proposed to the IAU committee around the same time that the discovery was announced. After the rumpus surrounding Sedna, the committee had promised that the decision-making process could be speeded up in exceptional cases.

For her part, Brown's wife Diane was not too keen on sneakily naming the new planet after their new-born daughter. What would happen if they had a son after a couple of years – would there be an eleventh planet available so that he, too, could be immortalized in the solar system? If not, couldn't that lead to jealousy, frustration or even a trauma? Diane thought that Mike could sometimes be very impulsive and thoughtless. A little over 3 months earlier she had been forced to persuade her husband to abandon the inauspicious idea of giving the large ice dwarf K50331A the nickname 'Dead Pope' (around the time that the object was discovered, the imminent death of Pope John Paul II dominated the international news). Fortunately, Brown had allowed himself to be convinced and settled for the less controversial nickname Easter Bunny. But he was not prepared to give Lila up so easily. Who knows, perhaps the IAU committee would support the proposal. There was no harm in asking.

In the meantime, Mike, Chad, and David had also agreed on how and in what order the discoveries of the three mini-planets would be announced. Because they had not yet conducted enough follow-up observations, Easter Bunny would have to wait for a while. Santa and Xena (alias Lila) would be announced about a month apart. After all, it is better to come forward with an important discovery twice than just once. The idea was to announce Santa to the world at the start of September. That would take the form of a talk and two poster presentations at the 37th annual meeting of the Division for Planetary Sciences (DPS) of the American Astronomical Society in Cambridge, England. Xena would then be announced sometime in October, preferably at a widely publicized press conference.

The news could in any case not be kept secret for much longer. A lot of his colleagues already knew that Brown was keeping one or more large ice dwarfs under his cap, and although he wrote FOR YOUR EYES ONLY on all his requests for follow-up observations, rumors were rife. It was therefore high time to bring everything into the open. But then in orderly phases, starting with Santa.

Keck Telescope photo of the two small moons of 'Santa' (Mike Brown/W.M. Keck Observatory)

On July 20, the abstracts of the three planned presentations were published on the website of the DPS conference in Cambridge. The summaries made it clear that a bright ice dwarf had been discovered at 7.8 billion kilometers from the Sun with a strange egg shape and a small moon. But there was of course no information on Santa's position – Brown and his colleagues wanted to keep the object to themselves for the time being.

The team also wanted the discovery of Santa to be the subject of a press conference during the meeting. That would put the hunt for large objects in the outer regions of the solar system back into the spotlight for a while, which would set the scene perfectly for the press conference on the discovery of Xena, a few weeks later. There could be no doubt about it – they were onto a real winner.

But before the month was out, the news had already leaked out. Mike Brown had not counted on a Spanish invasion.

Chapter 23
The Spanish Invasion

Pablo Santos-Sanz literally leapt into the air and gave a shriek of joy in his room at the Andalusian Astrophysics Institute in Granada. But then doubt set in immediately. Could it be true? Was it really possible that an astronomy student had discovered such a bright unknown object in the outer regions of the solar system? Shouldn't it have been found by someone else long ago? Was it really an ice dwarf or was it an asteroid that was much closer but which, for some reason, hardly seemed to move across the sky? Pablo thought maybe he shouldn't shout too soon.

The Observatorio de Sierra Nevada, from where the discovery of 2003 EL$_{61}$ was first announced (R. Romero, IAA-CSIC, courtesy Pablo Santos-Sanz)

It was the morning of Monday, July 25, 2005. Santos-Sanz studied the images made 2 years previously with a small 36-centimeter telescope at the Observatorio de Sierra Nevada, 2,900 meters up amid the ski slopes of Loma de Dílar in southern Spain. At the end of 2002, under the leadership of his study

G. Schilling, *The Hunt for Planet X*, DOI 10.1007/978-0-387-77805-1_23,
© Springer Science+Business Media, LLC 2009

supervisor José-Luis Ortiz, a photographic search for large ice dwarfs had been initiated at the observatory, but there had been some delays in analyzing the digital images from the spring of 2003 because of technical problems. Santos-Sanz, who had helped write the software to track down the moving points of light, was now trying to make up the backlog.

Discovery images of 2003 EL$_{61}$ (IAA-CSIC, courtesy Pablo Santos-Sanz)

And now he was sitting staring at the monitor with his mouth hanging open, eye to eye with a slowly moving speck of light in the constellation of Coma Berenices. It had been detected three times: on March 7, 9, and 10, 2003. And it was improbably bright for an ice dwarf. If it really were a little less than 8 billion kilometers away it was certainly the largest Kuiper Belt object ever found. This could be the most important discovery of Pablo's career. It was time to inform Ortiz.

Andalusian Catch

José-Luis Ortiz was awarded his PhD at the University of Granada in 1994 for his research into the cloud structure of the giant planets Jupiter and Saturn. In the summer of that year, comet Shoemaker-Levy 9 had had a spectacular collision with Jupiter. The icy fragments of the disrupted comet had drilled through the planet's upper cloud layers at high speed and evaporated suddenly and violently. As a postdoc in the group led by planetary scientist Glenn Orton of NASA's Jet Propulsion Laboratory in Pasadena, Ortiz was given the unique opportunity of analyzing the many images and measurements of the impact. After spending 2 years in California he returned to Spain in 1977, where he joined the staff of the Andalusian Astrophysics Institute.

Ortiz had been interested in ice dwarfs for some time and, together with his colleagues, he studied some of the large ones in detail, recording their dimensions, brightness variations, and rotation periods. And, using relatively small, automated telescopes and simple CCD cameras, he also set up his own search for large ice dwarfs. That had produced very little up to now but, with this spectacular discovery, Pablo's luck seemed to have taken a sudden turn for the better. The object was given the name OSNT11 (OSN stands for Observatorio de Sierra Nevada and the T for *trans*-Neptunian object) and, during the night of Wednesday July 27, Santos-Sanz reported his discovery to the Minor Planet Center in Cambridge, Massachusetts.

The Spanish team of José-Luis Ortiz. From left to right: Jaime Nomen, Reiner Stoss, José-Luis Ortiz, Pablo Santos-Sanz, Salvador Sánchez, Nicolás Morales, and Juan Rodríguez (Reiner Stoss, courtesy Pablo Santos-Sanz)

Three images of a slow-moving point of light is of course not much. More observations were required to make sure the object was indeed a distant ice dwarf. On July 28, Ortiz and Santos-Sanz therefore called in the help of the highly talented German amateur astronomer Reiner Stoss of the Starkenburg Observatory in Heppenheim. On the basis of the three images from March 2003, Stoss calculated roughly where the object should have been a few days later and found it on digital images in the archives of the Near-Earth Asteroid Tracking program. These additional positions were used to make new calculations and the ice dwarf – as it indeed proved to be – was also found on older images, eventually even on photographs from 1955.

It was then a small step to calculate where the mini-planet should be now. That evening, through an Internet connection, Stoss aimed a 30-centimeter telescope at an observatory on Majorca at the relevant part of the night sky and made 30 electronic exposures. These were sent directly to Ortiz and, before midnight, new emails were sent to the Minor Planet Center specifying the object's old archival positions and the results of the Majorca observations.

Brian Marsden, the director (or, according to some, the tsar) of the severely understaffed Minor Planet Center, had hardly devoted any attention to the email from Santos-Sanz the previous day. That wasn't so surprising: the images dated from March 2003, so the Spanish astronomers clearly didn't think they were that urgent. Furthermore there was no mention of a Kuiper Belt object and, as far as Marsden knew, the Andalusian Astrophysics Institute had little experience with observations of this nature. On the evening of July 27, Marsden felt he had better things to do.

But when he received the archival positions and the new measurements from Ortiz at the end of the afternoon of the 28, Marsden and his colleagues Dan Green and Gareth Williams started to sit up and take notice. And with good reason: this was clearly an exceptional object. After conducting a few quick internal checks and calculations, Williams announced the discovery just before 8.30 pm in Minor Planet Electronic Circular 2005-O36. The circular specified positions and orbital data so that other observers could also study the object. Because the original images dated from March 2003, it was given the provisional number 2003 EL_{61}.

Google

This was fantastic news for planetary research, but a slap in the face for Mike Brown. As he read about the discovery by Ortiz and Santos-Sanz that evening at his home in Pasadena he realized immediately that it was 'his' Santa: the brightness, the position – everything was identical. There went his scoop at the planetary science conference in England. And the honor of discovering one of the most bizarre inhabitants of the Kuiper Belt. From today, anyone with access to a large telescope could observe the object. The head start Brown and his colleagues had enjoyed until then would dissipate in a matter of months. It was their own fault of course: if Brown had reported the discovery of Santa to the Minor Planet Center immediately at the end of 2004, he would have gone down in history as its discoverer.

Mike Brown did not know José-Luis Ortiz and Pablo Santos-Sanz personally. Come to think of it, he could not remember ever coming across their names in connection with Kuiper Belt research. But the two Spaniards had nevertheless made an important discovery. Just after 7 pm Pasadena time, after he had got over his initial disappointment, Brown sent an email to Ortiz to congratulate him. As a prominent ice-dwarf hunter, it was the least he could do. In science, as elsewhere, you can sometimes be pipped at the post by your fellow researchers but they are, after all, still your colleagues.

Less than 20 minutes later, Brown received a concerned mail message from Brian Marsden. Ice dwarf expert Larry Wasserman at the Lowell Observatory in Flagstaff had also read the circular about the discovery of 2003 EL_{61} and immediately thought of the mysterious object K40506A mentioned in the Division for Planetary Sciences (DPS) abstracts issued by Brown, Trujillo, and Rabinowitz. They also referred to an exceptionally bright ice dwarf at 7.8 billion kilometers distance from the Sun. Surely there couldn't be so many bright ice dwarfs? Could it be the same object? And if it was, was it really an independent discovery or had the information on the position of K40506A have leaked out in some way or another?

Wasserman expressed his concerns to Gareth Williams, Williams discussed it later that evening with Brian Marsden, and Marsden sent an email to Brown. Was it possible that Ortiz and Santos-Sanz had somehow found out about the position of K40506A? After all, the abstracts for the DPS conference had been online for a week. Brown reacted with surprise. No, of course not. The abstracts did not specify the position and the name K40506A meant nothing to anyone except himself and his immediate colleagues. But it was indeed very coincidental that the Spanish researchers had discovered the object on 2-year-old images only a few days after the publication of the abstracts.

The rest of the evening, Mike Brown couldn't stop thinking about it. Had there been a leak inside his team? It was unthinkable. Could someone working at one of the observatories where they had studied Santa have let something slip? It was very unlikely. Or did the strange code they had always used provide more clues than they thought? On an impulse, Brown opened the homepage of the Google search engine and entered 'K40506A' in the search window. The ground seemed to be pulled away from beneath his feet. The search took him directly into the observation logbooks of a telescope at Kitt Peak Observatory in Arizona, which had been used for follow-up observations of Santa. Date, time, position, exposure time – it was all there for anyone to see on the Internet!

Although Brown had no proof, it looked as though the Spaniards had been guilty of foul play. If they were indeed looking for large ice dwarfs themselves, they would of course have been interested in the contributions of Brown and his colleagues to the DPS conference. And of course they would have read the team's abstracts on the conference website on or shortly after July 20. A simple Google search revealed a mine of information on the position of K40506A and, if the Sierra Nevada team themselves had old images of the constellation of Coma Berenices in their archives, it would be a piece of cake to calculate where this improbably bright ice dwarf was to be found. And then they brazenly announced their 'discovery', without even mentioning Brown's name once. It was theft, pure and simple. At least, it certainly looked that way.

Mike Brown with his daughter Lilah (Courtesy Mike Brown)

Brown didn't sleep a wink that night – and this time it was not because of his 3-week-old daughter Lilah, who regularly kept her parents awake at nights. Since she had been born, Mike had hardly been to the office, but now mother Diane had to manage on her own for a while. For the moment, the solar system was more important; daddy had to go to work. The following morning, on Friday, July 29, Mike nearly had a heart attack for the second time when he realized that that same Kitt Peak Telescope which had been used to study Santa had also been used for observations of Xena and Easter Bunny. That meant that their internal codes (K31021C and K50331A) were in the same observation logs. You didn't need to be Sherlock Holmes to realize that these objects were also two interesting ice dwarfs. If Ortiz and Santos-Sanz really were playing dirty, it wouldn't be long before they also announced these two 'discoveries.' It suddenly seemed very unwise to wait until October to officially announce the discovery of Xena and Easter Bunny.

Brown contacted Marsden, told him in confidence about the two other ice dwarfs he had discovered and expressed his concerns about the possibility that they could also be hijacked by someone else. His concerns proved well-founded, as Sebastian Hönig of the Max Planck Institute for Radio Astronomy in Bonn and Jean-Claude Pelle of the Southern Stars Observatory in Tahiti had already been in touch with the Minor Planet Center. Like a pair of eager private detectives, they informed Marsden and his colleagues that anyone could obtain information on at least three bright ice dwarfs via Brown's DPS abstracts.

Typing Error

There was only one thing for it: to announce the discovery of the other two objects as soon as possible. Brown provided the Minor Planet Center with the necessary information and around midday Gareth Williams issued two new Minor Planet Electronic Circulars. Xena was given the provisional number 2003 UB$_{313}$ and was announced at 12.33 pm Pasadena time in MPEC 2005-O41; Easter Bunny followed 17 minutes later in MPEC 2005-O42 and was given the provisional number 2005 FY$_9$. Later that evening, the Central Bureau for Astronomical Telegrams issued another official electronic circular from the International Astronomical Union (IAU, IAUC 8577), in which Dan Green described the discovery of the three ice dwarfs.

And then a press conference had to be organized – that too had to happen as soon as possible. Brown conducted frenetic consultations with Jane Platt, the press officer at NASA's Jet Propulsion Laboratory, which is operated by California Institute of Technology (Caltech). Platt thought it should be possible to set up a press conference at 4 o'clock that same afternoon in Pasadena. It would have to be a teleconference, where journalists could call in, because you couldn't expect many people to attend in person at such short notice. In the early afternoon, Brown called home to ask Diane to bring a razor and a suit as soon as possible. A press release was hastily put together with the suggestive title 'NASA-Funded Scientists Discover Tenth Planet'. And, because of the hurry, the release contained a typing error: it said that 2003 UB_{313} had first been recorded on October 31, 2003 instead of October 21.

The discovery of the tenth planet was therefore officially announced to the press at 4 pm on July 29, 2005. According to Brown, this was the worst conceivable time to release such important news: on a Friday afternoon in the middle of the summer holidays, and in a week in which NASA – 2.5 years after the fatal accident with the space shuttle Columbia – had once again found itself facing problems with the shuttle. This time, pieces of insulation foam from the fuel tank were breaking loose. Some of the more suspicious journalists suggested that the news about the tenth planet may have been released intentionally to divert media attention from the persistent problems with the space shuttle.

That attention was, of course, enormous. The telephone rang non-stop – on the West Coast, the discovery of Planet X received extensive coverage on television that evening and most papers opened with it the following morning. It was even on the front page of the *New York Times*, where science journalist Kenneth Chang, an old friend from Mike Brown's college days, managed to hold up the deadline long enough for the news to be included. What a day! Brown came home late, dead tired, and was only able to put all the events and emotions of the past 24 hours in perspective once he took little Lilah into his arms. Had Ortiz and Santos-Sanz really gleaned their information from the Kitt Peak observation logs? If so, why hadn't they also claimed the discovery of Xena and Easter Bunny, too? If you are breaking into a safe anyway, surely you would take the most valuable object? Or was the Spanish discovery of Santa really just pure coincidence? Why not? Surely astronomers couldn't be that dishonest?

Brown had already resolved earlier that Friday not to give in to suspicious thoughts. There was no hard evidence that the Spaniards had cheated, so you had to assume that their version of the story was correct and that it was just an unfortunate coincidence. He had even sent an email to Ortiz to inform him of the pending press conference and to explain why they were suddenly so keen to make the discoveries public. And he had again stressed that he did not suspect his Spanish colleagues of any foul play.

But not everyone took such a generous attitude. The media were of course quick to sense that something was going on. Brown had explained why he had decided not to wait any longer to make an announcement about the tenth planet, how – thanks to the discovery by Ortiz and Santos-Sanz – he had found out about the unintentionally accessible observation logs. And on the Minor Planet Mailing List (MPML), an international mailing list for professional astronomers and amateurs with a great interest in asteroids, rumors, suggestions, and suspicions were rife. Everything seemed to suggest scientific fraud and, no matter how correctly Brown continued to act, within a few days the situation had developed into a full-fledged row.

IP Addresses

The row focused largely on whether Ortiz and Santos-Sanz had taken a look at Brown's observation logs or not. The two Spanish astronomers were hardly available for comment and refused to be drawn into details when confronted with such questions. The discussion on the MPML quickly became very heated and, because neither of the two Spaniards subscribed to the list, the accusations of fraud were directed at Reiner Stoss, who had helped them calculate the orbit of 2003 EL$_{61}$ and who was an active member of the list.

Stoss had been at loggerheads with Brown before, or rather with Brian Marsden. When Brown announced the discovery of ice dwarf 2003 VB$_{12}$ in March 2004 and started to use the name Sedna for it on his own initiative, before it had been approved by the IAU, Stoss had protested against this violation of the usual procedures in a very original way. He had discovered a number of small asteroids himself, one of which had since been given a permanent number. As the discoverer, Stoss was entitled to give the chunk of rock a name and, in the spring of 2004, he submitted a proposal to the IAU committee, suggesting it be called Sedna, after the completely unknown German–American singer Katy Sedna. It seemed that the committee could hardly object to the proposal, which would make it impossible for 2003 VB$_{12}$ also to be called Sedna.

The whole affair ended not with a bang, but with a whimper (Marsden proposed to Stoss that he named his asteroid Katysedna – a suggestion that he never took up), but the entire asteroid world remembered that Stoss didn't hold the Minor Planet Center in very high regard and had little time for what he saw as Mike Brown's arrogance. Now that Ortiz and Santos-Sanz had been accused of plagiarism, Stoss of course defended them tooth and nail. He even added fuel to the fire by suggesting that Marsden had long known about Brown's three ice dwarfs and had intentionally withheld information from the astronomical community, which was blatantly contrary to the Minor Planet Center's mission.

Stoss, however, consistently avoided or simply ignored the simple question as to whether the two Spaniards had access to the positional measurements made by Brown and his colleagues before they announced their discovery. It was astronomer Richard Pogge of Ohio State University who provided the answer early in August. Pogge was the system administrator of Small and Moderate Aperture Research Telescope System (SMARTS) – a group of relatively small telescopes at Kitt Peak Observatory, including the one that Brown and his team had used to conduct their follow-up observations of Santa, Xena, and Easter Bunny. At the request of David Rabinowitz, Pogge took a look at the server logs of the SMARTS computers to see which web pages had been accessed when and by whom.

He struck lucky in no time. In the morning of Tuesday, July 26, at 10.08, 10.16 and 10.26 am Spanish time to be precise, the page containing the positional data for K40506A had been accessed three times by a computer with the

Internet Protocol address 161.111.165.49. It was then easy to determine that this address belonged to a computer at the Andalusian Astrophysics Institute. It proved to be the same computer that Pablo Santos-Sanz had used that evening to inform the Minor Planet Center of the discovery of OSNT11, the slowly moving point of light that had been found on the images from March 2003.

On Thursday, July 28, just before 11 am Spanish time, the SMARTS server was again accessed three times from Granada. This time the computer had the IP address 161.111.166.194, and it was used again 12 hours later by José-Luis Ortiz to inform the Minor Planet Center of the archival and new positions of the newly discovered ice dwarf. It was now clear beyond doubt: Brown's positions for Santa had been accessed before Santos-Sanz and Ortiz claimed their discovery. It was perfectly possible that they only started to examine their old images from 2003 on Tuesday afternoon and had therefore indeed committed astronomical plagiarism.

Mike Brown now started to have serious doubts too. A little late, in the eyes of his PhD students Kristina Barkume and Emily Schaller, who had been thinking for a while that he had become a little too kind and gullible since the birth of his first child. On August 9, Brown sent an email to Granada asking for an explanation, but received no reply. On August 14, at his wits' end, he finally submitted an official complaint to the IAU, with a request to instigate an internal inquiry.

Secrecy

That did not have much effect, however. Early in September, shortly after the DPS planetary science conference in Cambridge, England, Brown finally received an email message from Ortiz, in which he neither denied nor confirmed nosing around in Brown's observation logs. He did, however, criticize his Californian colleague for not reporting new discoveries to the Minor Planet Center immediately, but keeping them under his cap for some time. He claimed that this was not in the interests of scientific progress. Brown, he said, must realize that it was primarily because of his own strategy of secrecy that the two astronomers were now exchanging emails.

On September 16, through his Spanish colleague Jaime Nomen, Ortiz published a letter on the MPML in which he once again insisted on the importance of revealing new discoveries immediately. In the letter he admitted for the first time that the SMARTS logs had indeed been consulted and that they had been found by conducting a Google search for the codename K40506A from the DPS abstracts. But that search had been carried out a day after Santos-Sanz had discovered the object on the old Sierra Nevada images, and the only reason had been healthy scientific curiosity as to whether it might be the same body described by Brown and his colleagues in the abstracts. He added that there was no question of hacking, spying or violations of privacy since the SMARTS website was after all, accessible to everyone.

Brown of course saw things differently. The observation logs were never intended to be openly accessible – that was the result of a programming error, which had since been rectified. 'It's true that the information was available without braking into any sites,' he wrote on his personal website. 'It's also true that sometimes I don't lock the door to my house. I hope that people don't think it's therefore OK to come in and take my stuff.'

What they could not take away from him in any case was that he had discovered the tenth planet. 2003 UB_{313}, alias Xena, was the first ice dwarf larger than Pluto. Brown hoped that the IAU would reach a decision very soon about the name he had suggested for the mini-planet. He did not use the name in public. He had done that a year before with Sedna, and it had led to a wave of criticism. For the time being then, Planet X was still just 2003 UB_{313}, or he had to use the nickname Xena, but that would probably not be for much longer.

In reality, however, it was not that simple. The IAU has a committee that gives names to small solar system bodies, but as long as the official status of 2003 UB_{313} had not been determined, no one knew if the committee was authorized to make a ruling on its name. If Xena were to be classified as a real planet, there was a completely different set of rules and, according to tradition, it would be named after a Roman god. But if the discovery of large ice dwarfs in the Kuiper Belt meant that Pluto would be disqualified as a planet, then Xena was clearly not a planet either and it would be the responsibility of the IAU committee to give it a name. So before the object could be given a permanent name, agreement had to be reached about what exactly defines a planet.

Everyone knew by now that Brown was thinking of the originally Hindu name Lila. The webpage with all the background information on 2003 UB_{313} (www.gps.caltech.edu/~mbrown/planetlila) was of course given the name for very good reason. Brown had been working on it for several weeks and, on July 29, suddenly had to bring it online. In the rush, he forgot to change the name to something more neutral, which threatened to spark off another Sedna-like row. There seemed little point in changing the name, as that would just arouse more suspicion. Instead, Brown told everyone who asked that the page had been named after this new-born daughter. The missing 'h', he said, was a typing error. But if he was honest, he didn't seriously expect anyone to believe it.

José-Luis Ortiz appears to want to forget the soap opera around the discovery of 2003 EL_{61} and its two big brothers as soon as possible. In the spring of 2006, in any case, he claimed to be too busy for telephone interviews, and a visit to the institute in Granada by a foreign journalist was completely out of the question. Pablo Santos-Sanz, too, seemed uncooperative at first. He was only a student, he said, his English was pretty bad, and actually you should direct all your questions at his boss. With a little persuasion, he did admit that the Spanish team had perhaps not handled the whole business very well – a telephone call to Mike Brown on July 26 might have been appropriate. But there must be no doubt about the fact that, in the morning of Monday, July 25, he had come across that intriguing new object completely independently of

Brown. The Spaniards, he insisted, had done nothing wrong. If anyone had been at fault, it was Brown; he had kept his discoveries secret for many months.

Mike Brown has never met José-Luis Ortiz. But Pablo Santos-Sanz was one of the participants in an international workshop on *trans*-Neptunian objects held early in July 2006 in the Hotel Nettuno in Catania, Sicily. Brown and his Caltech colleagues were of course also present. When the Californian group discovered during the conference dinner in the hotel dining room that the good-looking guy with the dark curls two tables away was their Spanish colleague, they laughed and wondered if they ought to go over and introduce themselves.

The same evening, Mike and Pablo happened to meet in the elevator. Now they could no longer avoid each other. Okay, time to play the friendly American, Brown must have thought. He introduced himself, invited Santos-Sanz to drop by sometime in Pasadena and promised to organize a big party for him at his home. Pablo Santos-Sanz, who will go down in history as the official discoverer of one of the largest and most fascinating objects in the outer regions of the solar system, responded sheepishly. And not a word was said about the EL_{61} affair.

Chapter 24
Pas de deux

Robin Canup has difficulty keeping her hands still. On her computer monitor, a cosmic ballet is being played out before her eyes. There is no sound, but you can easily imagine music accompanying the dance. One minute it is subdued and serene, like an ethereal violin, and then it is dynamic and dramatic, with a lot of brass and percussion. Two dancers fall into each other's arms and a shower of sparks explode into surrounding space. Glowing fragments trace graceful paths across the three-dimensional stage, under the strict choreography of the force of gravity. Some of them merge to form a new ballerina, and the performance ends with an intimate *pas de deux*. Robin sits transfixed to the screen as the drama unfolds, her shoulders and wrists swaying slightly back and forth in the heavenly breeze, as though her slender body wishes to join in the dance.

Robin Canup in the ballet 'Paquita,' with Edward Stegge (Richard Mihran, courtesy Robin Canup)

G. Schilling, *The Hunt for Planet X*, DOI 10.1007/978-0-387-77805-1_24,
© Springer Science+Business Media, LLC 2009

You soon get too old for ballet shoes and tutus. For 8 years, Robin Canup was prima ballerina with the Boulder Ballet. In her office at the Southwest Research Institute, lying face down on a row of hefty-looking books on planetary research, there is a magnificent framed photograph, taken in 1997, showing her dancing in *Romeo and Juliet* to the music of Prokofiev, performed by the Boulder Philharmonic Orchestra. Three years later, she decided to call it a day. But if you have been doing ballet since you were 6 years old, it becomes a part of who you are. And watching Pluto's moon, Charon, forming from the dust of a cosmic collision is as elegant and dynamic as Stravinsky's Firebird or Tchaikovsky's Nutcracker Suite.

Canup's computer simulations of the birth of Charon offer the answer to a riddle that has been puzzling scientists for more than a quarter of a century. Charon is actually half as big as Pluto, and the two icy, rapidly rotating celestial bodies circle each other in a small orbit, always showing each other the same face. How did this remarkable 'double planet' come about? And what can their origin and history tell us about the early youth of the solar system? Interestingly, answers to these questions began to present themselves even before Charon was discovered.

The large satellites of the giant planets Jupiter and Saturn were formed during the birth of their mother planets, while the smaller ones are planetesimals of ice and rock, captured by the planets' gravitational pull. Mercury and Venus have no moons, and Mars once captured two asteroids which now orbit the red planet as the miniature satellites, Phobos and Deimos. All this we could understand and explain, but the Earth remained a problem. In relative terms, it has an exceptionally large moon and, up to the late 1960s, planetary scientists didn't really know why.

The samples of moon rock that the Apollo astronauts brought back to Earth didn't make matters any easier. Laboratory research revealed that the Earth and the Moon are of the same age and must have originated in the same part of the solar system. But the Moon contains much less iron than the Earth, and lower levels of certain volatile elements. Did the young Earth collide with another celestial body? With such a catastrophic impact, the intruder would have been pulverized, and its volatile components would evaporate. At the same time, part of the Earth's mantle would be ejected into space. This would be relatively light rock, as most of the Earth's iron is found at the planet's core. All those fragments could have come together to form the Moon.

This collision scenario was first put forward in the mid-1970s by William Hartmann and Don Davis of the Planetary Science Institute in Tucson, Arizona, and by Alastair Cameron and William Ward of the Harvard-Smithsonian Center for Astrophysics in Cambridge, Massachusetts. At the time, the theory did not receive much attention. People saw it as a far-fetched answer to a stubborn riddle. It was considered very unlikely that such massive collisions could occur after the formation of the planets.

Big Whack

It was not until 10 years later, when scientists had become used to the idea of large-scale cosmic impacts that interest in the 'Big Whack' theory started to gain momentum. At a conference on the origins of the Moon in Kona, Hawaii, in October 1984, planetary researchers concluded that the impact theory was actually the most likely explanation, although it was certainly not clear as yet how the fragments created by such an impact could merge together so quickly to make a new celestial body.

Robin Canup at her desk at the Southwest Research Institute in Boulder (Courtesy Robin Canup)

It was Robin Canup's computer simulations that finally offered a solution. At the University of Colorado, together with Larry Esposito, Canup had conducted theoretical research into the formation of mini-moons within Saturn's rings, caused by particles in the rings accreting and aggregating together. They presented their results on this topic at a planetary science conference in 1993 and after their talk William Hartmann, who had been in the audience, asked Robin to apply the same methods to the origins of the Moon. After much time and effort, she succeeded and, on September 25, 1997, she published her findings in *Nature*, together with Shigeru Ida and Glen Stewart. That was the same year in which the beautiful photograph of her dancing in *Romeo and Juliet* in the Macky Auditorium in Boulder had been taken.

The simulations constructed by Ida, Canup, and Stewart showed that the fragments caused by a planetary collision could merge together in less than a year to form a celestial body the size of the Moon. The exact outcome depends on the original distribution of the fragments, and therefore on the nature of the impact itself. It might result in one large moon, or two of comparable size. There might even be one large moon and a lot of smaller ones but, according to the three astronomers, the smaller ones would eventually – in a matter of a million years or so – be 'swept up' by the large one.

This was Robin's fourth scientific publication and the first in the prestigious weekly *Nature*. The article generated an enormous wave of publicity and the collision theory for the origin of the Moon now became widely known outside astronomical circles. As did Robin – after all, it's not every day that a prima ballerina explains the birth of the Moon.

Robin Canup grew up in Poughkeepsie, a small town north of New York, where she showed a great interest in science at school, danced to her heart's content at the Poughkeepsie Ballet Theatre, and watched Carl Sagan's TV series *Cosmos* with an almost religious devotion. After completing her degree in physics at Duke University in North Carolina with a *magna cum laude,* she moved to Boulder, where she is now the director of the Southwest Research Institute and her desk is covered in ring integrals and differential equations.

Canup's interest in the origins of the Moon has never waned. On August 16, 2001, together with Erik Asphaug of the University of California in Santa Cruz, she once again published extremely detailed computer simulations in *Nature*. These latest results suggest that the Big Whack occurred a little later than was first assumed, and that the young Earth must have collided with a planet the size of Mars.

Bill McKinnon, who first demonstrated that Charon must have been born in a catastrophic collision (Photo archive of the Division for Planetary Sciences, AAS)

A few years later, Canup turned her attention to Pluto and Charon, the other 'double planet' in the solar system. Bill McKinnon of Washington University in St. Louis had already demonstrated in 1984 that the properties of Pluto and Charon could only be explained by assuming that the couple were formed in the aftermath of a catastrophic impact, just like the Earth–Moon system. But credible simulations had never been carried out. So it was high time, especially because Pluto and Charon were no longer the only binary ice dwarfs. Could all the other binary companions and moonlets in the Kuiper Belt also have been caused by impacts?

Icy Duos

The first ice dwarf satellite was found by chance at the end of 2000. On December 22 and 23 of that year, Christian Veillet and his colleagues at the Canada–France–Hawaii Telescope in Hawaii took photographs of the Kuiper Belt object 1998 WW_{31}. They then discovered that it was actually two large objects of almost equal size, circling each other in a relatively wide orbit. On photos made a year previously by other astronomers the companion was also just about visible, but no one had noticed it. Veillet and his colleagues announced their discovery on April 16, 2001, and from that moment the hunt for ice dwarf satellites was on. At the end of that year, the Hubble Space Telescope discovered a second pair, which were recently given the Gnostic names Logos and Zoe. Since then, several dozen binary Kuiper Belt objects have been discovered, mainly by Keith Noll of the Space Telescope Science Institute and Denise Stephens of Johns Hopkins University using the Hubble Space Telescope. The actual number probably runs into the hundreds.

Could all of these satellites and binary objects have been caused by cosmic collisions? Robin Canup decided to test that theory first on the best-known couple: Pluto and Charon. It must be possible to use the same software she had used with Erik Asphaug for the successful simulations of the creation of the Moon to model the birth of Charon. The idea was actually very simple: you let the computer run a few hundred different collision scenarios and see if any of the results display similarities with the Pluto–Charon system.

Each simulation takes a few days on an average computer, so in the summer of 2004 Canup set 16 computers to work at the same time. Then she wrote software to visualize the results of the simulations: the enormous flow of figures and data had to be converted into short films showing how two celestial bodies collide, how the resulting fragments are scattered throughout space, and what eventually emerges as an end product.

Sometimes the colliding bodies were equal in size, sometimes they were different. Sometimes they collided head-on, and sometimes they only gave each other a glancing blow. Canup was also able to vary the velocity and internal structure of the two colliding objects in any way she liked. In this way, she built up her own private collection of alternative outcomes.

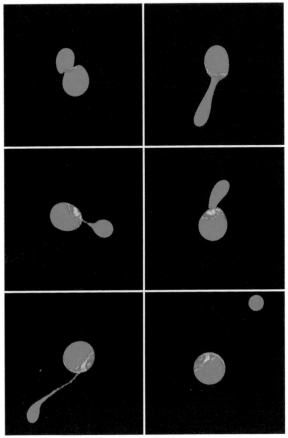

Scenes from one of Robin Canup's computer simulations of the formation of Charon as a result of a collision with Pluto (Courtesy Robin Canup)

All the individual collisions have their own style, dynamics, and rhythm but, in the films, each one is as graceful as a ballet performance, and they all obey the laws of the same strict choreographer: gravity. And to Canup's surprise, some of these ice ballets resulted in a planet with a large, fully formed moon as early as 1 day after the collision. This was a completely different story to the Earth–Moon system, where the fragments of the collision only merged together to form the Moon after some time. The collision theory appeared to explain everything: the rapid rotation of Pluto and Charon, the small orbit, and their relative sizes.

Confirmation

The fact that there was a lot of activity in the Kuiper Belt in the distant past is no longer only suggested by Canup's computer calculations, but is supported by the amazing observations of Mike Brown and his colleagues. On Friday,

January 28, 2005, which happened to be the same day that Canup published her Charon simulations in *Science*, Brown pointed the 10-meter Keck Telescope at Santa, the large, bright ice dwarf he had discovered a month previously. Close to Santa, there was an extremely dim point of light, a small satellite about 300 kilometers across. In addition, periodical variations in brightness showed that Santa rotated on its axis at incredible speed, in less than 4 hours. This fast spin was partly the reason for Santa's extremely elongated shape.

These discoveries are a shining confirmation of the collision theory. Rapid rotation could very probably be caused by a cosmic collision giving the object an enormous wallop and sending it spinning. Rudolph, as the tiny satellite was nicknamed, could be one of the fragments left over from the collision. And, indeed, spectroscopic observations show that it is just a large lump of ice, the same material as Santa's mantle.

On June 30, 2005, the Keck Telescope revealed a *second* icy satellite close to Santa. This one was given the name Blitzen (like Rudolph, named after one of Santa's reindeer). Blitzen is probably not much larger than 170 kilometers across and has a slightly smaller orbit than Rudolph. The two orbits are at a steep angle to each other, which also suggests that Santa's satellites are fragments left over from a collision. The discovery of Blitzen was not made public until late November, long after Santa had been given the provisional designation $2003\ EL_{61}$.

The orbital properties of the two small moons enabled the mass of $2003\ EL_{61}$ to be calculated and, because the dimensions of the elongated celestial body are also known (roughly $2000 \times 1500 \times 1000$ kilometers), it was possible to determine its density. This turns out to be around 3 grams per cubic centimeter – much higher than that of ice. The ice mantle therefore cannot be very thick, and $2003\ EL_{61}$ must consist largely of heavier rock. That too fits in neatly with the collision scenario: $2003\ EL_{61}$ was originally much larger, but most of the thick icy mantle was shattered and blown away by the collision. The two tiny satellites are the silent witnesses of that primordial clash.

In the meantime, on March 15, 2007, Brown and three of his California Institute of Technology (Caltech) PhDs announced in *Nature* that they had found more fragments from the same collision. Not in orbit around $2003\ EL_{61}$, but in their own orbit around the Sun. These fragments are no longer in the direct vicinity of $2003\ EL_{61}$, but they do circle the Sun in similar paths, with the same orbital period, inclination, and eccentricity. That suggests that they have a common origin. In addition, like Rudolph and Blitzen, they all seem to consist of pure ice. That is very exceptional, as the surfaces of most ice dwarves are contaminated by other material.

Incidentally, on September 10, 2005, the Keck Telescope also revealed a satellite about 400 kilometers in diameter close to $2003\ UB_{313}$, alias Xena, the 'tenth planet.' Brown did not take long to think of a suitable nickname for Xena's moon. What better name than Gabrielle, after the Warrior Princess's inseparable companion? The discovery of Gabrielle was announced on September 30. And in February 2007, the International Astronomical Union published Brown's discovery of small satellites orbiting the two large Kuiper Belt objects

Orcus and Quaoar, which were first photographed by the Hubble Space Telescope in November 2005 and February 2006.

2003 UB$_{313}$ ('Xena') and its satellite 'Gabrielle,' as observed by the 10-meter Keck Telescope
on Hawaii (W.M. Keck Observatory)

Robin Canup's convincing computer simulations and Mike Brown's intriguing discoveries leave no doubt at all that, long ago, the Kuiper Belt was a
cosmic crash-test course. Celestial bodies hundreds of kilometers across
smashed into each other and fragments flew off in all directions, resulting in a
wide variety of companions. Canup has since succeeded in explaining the
current orbits of Pluto's two small satellites Nix and Hydra, by assuming they
once experienced orbital resonances with Charon.

That turbulent period in the history of the solar system took place, however,
a few billion years ago, when the Kuiper Belt was much more densely populated
than it is now. We shall never see these catastrophic cosmic collisions themselves, only what they left in their wake. The silent remains of past activity, like
abandoned props gathering dust on the stage after a dazzling ballet performance. Would it be possible for us to study this still life and reconstruct all the
graceful steps and pirouettes that were played out on this stage all those billions
of years ago?

Chapter 25
Model from Nice

Hal Levison doesn't like observing. A little inconvenient for an astronomer, you might think. Unless of course you are a theorist. Then there is enough for you to study – not in the night sky, but on your computer screen. And in your mind, of course. Not just one solar system, but hundreds, all slightly different. Not one universe, but as many as you might want. And Levison can do all this thinking anywhere he likes. Not only in the middle of the night, on a cold mountain top, but during the daytime, in his office. Or on the Cote d'Azur. 'My work consists largely of thinking,' said Inspector Morse, the prótagonist of a British detective series, to his sidekick sergeant Lewis when the latter expressed surprise that his boss was not at his desk. 'And today I choose to do my thinking on a bench in the park.' And that's just how it is. Contemplating, discussing, suddenly getting brilliant ideas – you can do all that just as well on the French Riviera over a glass of white wine. Perhaps even better.

So Levison is not at all jealous of colleagues who pass their nights at the telescope. It messes up your daily rhythm, and you have to spend a lot of time away from your wife and kids, sitting all alone on some remote mountain. Until recently, at the smaller observatories, you also had to keep a close eye on telescope to ensure that it precisely tracked the motion of the night sky – a soul-destroying chore. If you're lucky, you can get disgusting coffee and pick up one radio station – and then have to sit and listen to the same seven songs all week. Observations are very important, but Levison is happy to leave them to someone else.

Not that the universe holds no fascination for him. On the contrary, in the early 1970s, as the Apollo program was coming to an end, 12-year-old Harold knew that he wanted to become an astronaut. He devoured Carl Sagan's *The Cosmic Connection*, was inspired by his physics teacher Scott Negley, and even ground his own telescope mirrors. But even then he enjoyed building telescopes more than looking through them. Precision mechanics, electronics, messing around with computers, writing software – they were more Hal's thing. His first publication did not appear in an astronomical journal but in *Byte*. And during his Masters course at the University of Michigan he worked on the construction of what would later become the 2.4-meter Hiltner Telescope on Kitt Peak in Arizona.

It was therefore not so strange that, in Ann Arbor, Levison became fascinated by the work of Douglas Richstone: stellar dynamics or, in other words, theoretical

G. Schilling, *The Hunt for Planet X*, DOI 10.1007/978-0-387-77805-1_25,
© Springer Science+Business Media, LLC 2009

research into the motions of the billions of stars in a galaxy. He could indulge his passion for programming, and within the most fascinating area of science you could imagine. In 1986 Levison was awarded his PhD for theoretical study of the dynamics of elliptical galaxies. He would never have believed that he would end up doing solar system research. What could possibly be interesting about that?

But it's strange how things can turn out sometimes. Hal Levison, who has been working at the Southwest Research Institute in Boulder, Colorado, since the end of 1992, is now an authority on the dynamics of the Kuiper Belt. Together with Canadian theorist Martin Duncan he developed the Swift software package, with which you can conduct gigantic computer calculations of the mutual gravitational effects of millions of test particles. These simulations can provide an insight into the early evolution of the solar system. Looking back in time – now that's what Levison calls observation. And if you can do it in the South of France, even better.

Levison owes his 'French connection' to his colleague Alessandro Morbidelli – an Italian mathematician and astronomer who has been working at the Observatoire de la Côte d'Azur in Nice since 1992. Like Hal, Alessandro felt the call of the cosmos at an early age. He was an active member of an amateur astronomy club, built his own 25-centimeter telescope, and wrote articles about observing variable stars in the Italian monthly *l'Astronomia*. Of course, he wanted to become an astronomer, but that wish only came true when he came to Nice, after he had been awarded his PhD in Namur, Belgium, for research into a complex mathematical issue in chaos theory. His knack for math proved very useful on the Riviera, where he devoted himself entirely to solar system dynamics.

Like Levison, Morbidelli has never conducted professional observations, except for the time he and colleague Brett Gladman did follow-up measurements on the ice dwarf 1998 WW_{31} and failed to notice that it actually consisted of two objects. That was only discovered a year later by Christian Veillet. But because of his background as an amateur astronomer, he knew the night sky better than many professionals. And during his many hiking expeditions – Morbidelli is an active mountaineer and has climbed many peaks, including Mont Blanc – he regularly directs his gaze up to the stars.

Strange Pair

It is intriguing that, without using a telescope or any other optical aid, Morbidelli can see the cosmic scars, the causes of which he and Levison have explained. But these are scars that everyone can see. In fact, which everyone *has* seen. We are all familiar with the Man in the Moon. The dark patches on our nearest cosmic neighbor are closely connected to the Kuiper Belt, way out beyond the orbit of the distant planet Neptune. Mare Tranquillitatis (the Sea of Tranquility), where Neil Armstrong left his footprints in July 1969, was almost certainly created by the impact of a stray ice dwarf.

At the end of the 1990s, Morbidelli worked regularly with Levison's colleague Bill Bottke and visited Boulder once or twice. Hal got on really well with Morby (as Alessandro is mostly called by his friends and colleagues) and before long he came up with the idea of spending a sabbatical at the Observatoire de la Côte d'Azur. Hal's wife Sarah had always wanted to live in France, and their daughters Rachel and Shaina were still young. In 2002 the Levison family boarded a plane that headed for Nice. And the move led to an exceptionally productive partnership with Morbidelli. Since then, Hal spends at least 1 month a year on the Riviera.

Hal Levison (Photo archive of the Division for Planetary Sciences, AAS)

Alessandro Morbidelli (Photo archive of the Division for Planetary Sciences, AAS)

They are a strange pair. Hal Levison is a giant of a man, big and heavy, with a ruddy face, a wild beard, and a ponytail. Alessandro Morbidelli is 7 years younger, lean and wiry, with glasses and a constant smile on his face. You would more likely expect to see Levison riding a Harley Davidson with a bandana tied around his head, than with his nose in *The Astronomical Journal*; Morby is more the archetypical scientist and computer nerd. But when it comes to solar system dynamics, they have certainly found one another and in Nice, step-by-step, a successful model was gradually pieced together which has since been used to solve countless cosmic mysteries.

The Nice model was developed in close cooperation with Brazilian astronomer Rodney Gomes and Morby's Greek postdoc Kleomenis Tsiganis. The model made a grand entrance in three articles in *Nature* on May 26, 2005, and since then its popularity has only increased. Not surprising really, since the model explains the orbital characteristics of the giant planets, the existence of Trojans (asteroids which go around the Sun in a similar orbit to Jupiter), and the mysterious timing of the cosmic bombardment that created the dark 'seas' on our Moon. Levison and Morbidelli think that they can also use the model to explain the current structure of the Kuiper Belt.

It may not have been dreamed up on a bench in the park, but it is certainly a brilliant piece of detective work. Like cosmic counterparts of Inspector Morse, Levison, Morbidelli, Gomes, and Tsiganis had to solve a celestial crime of violence on the basis of difficult-to-find clues, half wiped-out tracks, shreds of evidence, and dubious alibis. The fact that violence had been committed in the heavens was beyond dispute. It had been widespread and very severe, and was caused by stray giant planets and scattered lumps of ice.

Migration and Resonance

It was of course nothing new that the giant planets had gone a little astray in the early years of the solar system. It was inevitable, given their mutual gravitational interaction with the bodies surrounding them. First it was caused by the remains of the rotating gas disk from which the planets were formed, and later by the countless left-over planetesimals which populated the outer regions of the solar system – the first products of the accretion process that led to the formation of the planets.

Planetary migration already played a central role in the work of theorists like Julio Fernández and Renu Malhotra. But one mystery had not yet been solved satisfactorily. All the models predict that the giant planets will eventually settle into nice, circular orbits around the Sun, all in exactly the same plane. In reality, however, the orbits of Jupiter, Saturn, and Uranus are fairly eccentric (the distance from Saturn to the Sun, for example, varies between 1.35 and 1.5 billion kilometers) and the orbits of the two outer giants are at an angle of 1 or 2 degrees to that of Jupiter.

It was Kleomenis Tsiganis who came up with a solution that was both brilliant and unorthodox. Imagine that all four of the giant planets originally orbited the Sun a lot closer to each other – all within the current orbit of Uranus. They would then have had a much shorter orbital period. Saturn would perhaps have completed an orbit in less than 20 years, instead of its current 29.5 years. As it migrated outwards, there must have been a moment at which Saturn's orbital period was exactly twice that of Jupiter. That orbital resonance would have resulted in periodical and mutually reinforcing gravitational disturbances, causing the orbital inclination and eccentricity of the two planets to increase considerably.

It sounds good, but how do we know whether it is true? That's where Levison's Swift program comes in. You just enter the initial parameters, press the start button, and see how your model solar system evolves. Then do it again and again, with slightly different parameters, and try to identify patterns in the results. The results of the computer simulations were surprising. The orbits of Jupiter and Saturn did indeed end up with the correct inclination and eccentricity, but the resonance between the two giant planets also proved to have a substantial impact on Uranus and Neptune. The latter two planets moved around in chaotic orbits for millions of years, finally coming more or less to rest at a much greater distance from the Sun. In about half of the simulations, Uranus and Neptune even switched position: the planet that started off closest to the Sun ended up in the furthest orbit, and vice versa.

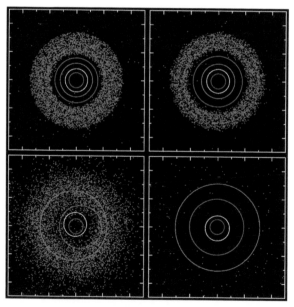

Scenes from a computer simulation of the solar system's early evolution according to the Nice model. Planets migrate, planetesimals are scattered away, and Uranus and Neptune swap orbits (Courtesy Alessandro Morbidelli)

It all sounded very promising, but at first there seemed to be one problem. Jupiter shares its orbit with many tens of thousands of large and smaller asteroids, the Trojans. Although they have a wide variety of orbital inclinations and eccentricities, about half of them move along a few hundred million kilometers ahead of Jupiter, while the other half follow along behind the planet at a comparable distance. It is quite conceivable that a large number of rocky fragments clustered together in these two more or less stable segments of Jupiter's orbit in the early stages of the solar system, but during the period of orbital resonance with Saturn that stability must have been disturbed and Jupiter would have lost all of its Trojans.

At first, the presence of the Trojans seemed to completely invalidate the migration and resonance scenario. Until Morbidelli suddenly had a brilliant idea. Sure, the two Trojan zones were instable during the period of orbital resonance with Saturn, so that the asteroids present in them at the time slowly but surely disappeared. But that instability also meant that other objects could enter and leave these areas unhindered. During this period of instability, therefore, small bodies were entering and leaving the Trojan zones all the time. In other words, they only inhabited the zones temporarily. As Saturn then continued to move further outwards, the orbital resonance was lost and the Trojan zones retrieved their stability. The bodies they contained at that time were no longer able to escape and stayed there permanently.

It seemed to be the answer to the whole problem, but Levison had little confidence in it at first. 'If you're right, I'll buy you dinner,' he told Morbidelli before leaving for a short holiday with his wife and kids in Avignon. When he came back, he found computer printouts and plots on his desk with convincing proof. Morby had conducted the necessary simulations and the predictions tallied almost completely with reality. According to the calculations, around a thousandth of a percent of all planetesimals in the outer regions of the solar system would eventually be captured in one of the two Trojan zones. Their orbital properties would be identical to those of the real Trojans and their predicted total mass – roughly a millionth of that of the Earth – came reasonably close. Morbidelli had a delicious meal.

The computer simulations also predicted that the Trojans would be more like porous, icy cometary nuclei than rocky asteroids. After all, according to the calculations, the captured objects came from the outer regions of the solar system, where their orbits were disturbed by the effects of Neptune's gravity. This prediction seems to have been confirmed by observations of the Trojan Patroclus. Patroclus was found to have a relatively large companion, Menoetius, which enabled its mass and density to be determined. The latter proved to be less than 1 gram per cubic centimeter, which suggests that Patroclus and its companion consist largely of ice.

One of the final predictions of the model is that there must also be Trojans in Neptune's orbit. According to the calculations, however, they should be much more widely dispersed, in terms of both the shape and orientation of their orbit. In August 2001 astronomers at Lowell Observatory had indeed discovered the first Neptunian Trojan, and four more have since been found, one of which has an orbital inclination of no less than 25 degrees. In reality, there are almost certainly thousands of Trojans in Neptune's orbit – resoundingly confirming the theory once again.

Bombardment

But the best was yet to come. The Nice model proved to offer a conclusive explanation not only for the orbital properties of the giant planets and the existence of icy Trojans in the orbits of Jupiter and Neptune, but also for the dark patches on our own Moon. The lunar 'seas' – which in reality do not contain a drop of water – are colossal impact basins, which have filled up with basalt lava from the Moon's interior. They therefore mark the locations where, billions of years ago, the Moon was struck by large cosmic projectiles. This cosmic bombardment is not so surprising: there must have been a lot of rubble in the inner solar system shortly after the Earth and the Moon were formed. The only strange thing is that it took place some 700 million years too late.

The Earth and the other planets are about 4.6 billion years old. The Moon was probably the result of a collision between the recently formed Earth and a

Impact basins on the moon (visible as dark 'lunar seas' close to the left edge of the moon in this Galileo image) were formed during the 'Late Heavy Bombardment' (NASA/JPL)

protoplanet the size of Mars, but that cannot have happened much later. Studies of the moon rock brought back to Earth by the Apollo astronauts show, however, that the impact basins on the Moon are only 3.9 billion years old. It is as though, after a tumultuous start and a relatively calm period of about 700 million years, all hell suddenly broke loose in the solar system.

This time it was Morbidelli's Brazilian colleague Rodney Gomes who pretty much got the ball rolling in the search for an answer. Suppose the migration and resonance model was largely correct and the outer three giant planets originally moved in much smaller orbits around the Sun? Then, said Gomes, it must be possible to set the initial conditions such that the orbital resonance between Jupiter and Saturn occurred 700 million years after the planets were formed. The resulting chaos in the outer regions of the solar system would then have led to an enormous peak in cosmic impacts. That would explain the 'Late Heavy Bombardment' of the Moon.

Levison's Swift software was once again used to test this idea against 'reality.' And again all the pieces fitted together perfectly. The resonance between Jupiter and Saturn generated enormous disturbances in the orbits of Uranus and Neptune, giving the Kuiper Belt beyond the orbit of Neptune a thorough 'shake-up.' Ice dwarfs were scattered in all directions and, within a period of some tens of millions of years a total of some 10 quintillion kilograms of cometary material impacted on the Moon. This undoubtedly included a large number of small cometary nuclei, but most certainly also a few large ones, which were responsible for the lunar 'seas.'

Of course, this explosion of violence in the cosmos not only affected the Moon. It is just that only half of the surface of the small planet Mercury has been charted and not yet accurately dated, while Venus, the Earth, and Mars have experienced so much geological activity in the past few billion years that the scars of the original bombardment have largely been wiped out. Which is not to say that no evidence of the primeval comet shower can be found on the Earth today. Because of its large gravitational pull, ten times as much cosmic ice fell onto our planet than on the Moon, as a result of which some 6% of all water on the Earth today probably comes from molten comets.

And it was not only lumps of ice that were scattered throughout the inner solar system. The same orbital resonances which caused so much chaos in the Kuiper Belt also wreaked havoc in the asteroid belt, which consists primarily of rocky planetesimals. The resulting asteroid bombardment probably lasted considerably longer (perhaps more than a hundred million years); it is therefore quite possible that some of the large impact basins on the Moon were caused not by ice dwarfs but by asteroids. Unfortunately, the Nice model cannot provide specific details on the relative share in the bombardment of these two types of objects, nor over how long it lasted. To find answers to these questions calls for further study of the soil samples brought back from the Moon.

Whether the Nice model can also explain all the properties of the Kuiper Belt in detail remains to be seen. Renu Malhotra has already shown that Neptune's outward migration led to the formation of plutinos – ice dwarfs which, like

Pluto, have an orbital resonance with Neptune (see Chapter 19). But the existence of ice dwarfs in extremely elongated and highly inclined orbits ('scattered disk' objects) and the relatively sharp outer edge of the Kuiper Belt at 7.5 billion kilometers from the Sun have not yet been explained. At least, not in a way that is convincing and quantifiable.

Levison and Morbidelli are also still working on a satisfactory explanation for the mass exodus from the Kuiper Belt. According to the Nice model, it must have contained a thousand times more material than it does now, otherwise there would not have been sufficient interaction with the giant planets to force them into large-scale orbital migration. The computer simulations work best when the original amount of material is assumed to be 35 Earth masses. Without such a high initial mass, the formation of large ice dwarfs like Eris and Pluto cannot be properly explained and there would be insufficient collisions in the Kuiper Belt to account for the large number of binary objects and ice-dwarf satellites.

But if you put all the ice dwarfs together on a set of scales, they would not amount to much more than around 4% of the mass of the Earth. In other words, the Kuiper Belt must have lost about 99.9% of its original inhabitants at some point. It is very easy to attribute that exodus to gravitational interactions caused by the chaotic wanderings of Neptune and then to link it to the Late Heavy Bombardment, the creation of Trojans, centaurs, and scattered disk objects and the formation of the Oort Cloud. But providing a solid, theoretical foundation for the idea is something else.

Hal Levison is very much aware that a theory, no matter how successful it proves to be, is not the same as an observed fact. Anyone who observes a new ice dwarf need not be afraid that it will no longer exist the following week. But one indisputable error can mean the end of the road for the Nice model, despite the fact that it now holds all the best cards. And if the model proves unable to provide a satisfactory explanation for observed reality, it will be time to seek an alternative.

At the same time, that is the charm of this line of work. Suppose someone found the definite answer. Then there would be no mysteries left to solve. And then what would you do out there on your park bench or on a sun-drenched pavement cafe on the Côte d'Azur?

Chapter 26
Planet Under Siege

There was nothing Brian Marsden could do about it. If it were up to him, he would give 2003 UB$_{313}$ an official name as soon as possible. It was clear to everyone that the discovery of this gigantic ice dwarf, announced by Mike Brown on July 29, 2005, was a historical event which would generate enormous publicity. The year before, the International Astronomical Union (IAU) had introduced a special emergency procedure for exceptional cases like this. The IAU's Committee on Small Body Nomenclature could now in theory decide on a definite name for a new object within a day. But, in the case of 2003 UB$_{313}$, everyone first had to be in agreement about its status – was it yet another ice dwarf, or was it really a new planet? If it was indeed a planet, Marsden's committee had no say in what it was to be called.

Of course, 2003 UB$_{313}$ – known for more than a year by its nickname Xena – was welcomed by media and public alike as the tenth planet. And not in the last instance by Mike Brown himself. After all, there had been speculation about a planet somewhere beyond the orbit of Pluto for decades and Xena was substantially larger than the remote ice planet. There was no doubt about it: the discovery of Planet X was a fact.

Unless of course Pluto itself was not a planet at all, just an overgrown ice dwarf. Not a full member of the planet family, but one of the largest inhabitants of the Kuiper Belt. Because then Xena would of course not be a planet either. So what were the criteria for qualifying as a planet? And who actually decided that?

Pluto's status as a planet had been a matter of debate since 1930. Clyde Tombaugh may have discovered the remote body during his search for the hypothetical Planet X, but it was clear from the start that Pluto was much smaller than expected. What is more, it moved in an extremely elongated and highly inclined orbit which, for a period of 20 years in each orbital period, brought it closer to the Sun than Neptune. Partly due to a successful publicity campaign by Lowell Observatory, little Pluto went down in history as the ninth planet in the solar system, yet astronomy books still described it as an odd man out.

And things didn't get any better as time passed. Although official astronomical handbooks stated a quarter of a century after Pluto's discovery that it was only 10% less massive than the Earth, these estimates were regularly adjusted

G. Schilling, *The Hunt for Planet X*, DOI 10.1007/978-0-387-77805-1_26,
© Springer Science+Business Media, LLC 2009

downward. The final blow came in 1978, when Jim Christy discovered Pluto's large satellite, Charon. The diameter and mass of the dwarf planet could at last be calculated. It proved to be a ball of ice only 2,300 kilometers across and no more than 0.2% of the mass of the Earth. It was the runt of the litter.

In theory, of course, there is no reason why an icy little runt in an oblique orbit around the Sun should not be called a planet. But it is a little more difficult to justify if hundreds more of them are found. Does that mean that all the Kuiper Belt objects are suddenly planets, too? If not, where do you draw the line? Or is it perhaps time to look reality in the face and strip Pluto of its planetary status?

Demotion

Brian Marsden would in any case not shed a tear if Pluto were to be demoted. He was in no doubt at all that it was one of countless ice dwarfs in the Kuiper Belt. It had been a mistake to call it a planet in 1930 and the sooner that could be put right the better. But not everyone looked at it that rationally. A lot of people, even (or perhaps especially) scientists seemed to attach a lot of emotional value to Pluto's status as a planet.

Marc Buie, Bob Millis, and Larry Wasserman of Lowell Observatory – where Pluto had been discovered in 1930 – naturally defended its status tooth and nail, even when their own Deep Ecliptic Survey unearthed one new ice dwarf after another. And of course Clyde Tombaugh, who kept abreast of all new discoveries in the outer regions of the solar system deep into old age,

Brian Marsden in his office at the Harvard-Smithsonian Center for Astrophysics (Harold Dorwin, Harvard-Smithsonian Center for Astrophysics)

continued to believe steadfastly until his death on January 17, 1997 that 'his' Pluto was a genuine planet. A lot of other American astronomers felt the same way – after all, Pluto was the only planet not to have been discovered by a European.

When the general public got wind of the plans to demote Pluto, emotions also started to run high. Hadn't we all learned at school that there were nine planets? You can't just go changing things like that – what on earth were these astronomers thinking? It was as though a handful of unworldly mathematicians suddenly announced that two and two now made five. Marsden might well claim that Pluto was not being demoted, that it would now be the giant of the Kuiper Belt rather than the midget of the planetary system, but no one was convinced.

It was a very sensitive issue among planetary scientists too. Alan Stern of the Southwest Research Institute in Boulder, for example, had been lobbying NASA for years to obtain funding for an unmanned space flight to Pluto – the only planet which had never been studied by a space probe. If Pluto were suddenly to be stripped of its planet status, the project would probably never get off the ground.

Ironically, it was the revolutionary observations of Plutophiles Alan Stern and Marc Buie that first gave rise to the widespread public commotion about Pluto's status. Using the European Faint Object Camera on board the Hubble Space Telescope, Stern and Buie had taken photographs of Pluto and, on the basis of the images, composed a rough 'world map' of the remote body. The map was revealed to the world on March 8, 1996 at a press conference at NASA Headquarters in Washington, D.C.

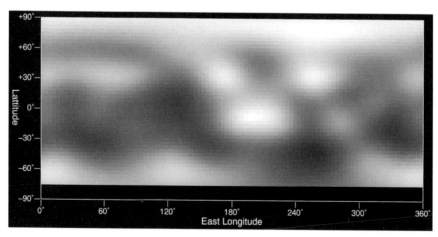

A map of Pluto, created on the basis of Hubble Space Telescope observations (A. Stern (SwRI), M. Buie (Lowell Observatory), NASA, ESA)

But a few days earlier, reporter Dan Vergano of *USA Today* had taken advantage of the imminent publication of the Hubble photographs to draw attention once again to the debate on Pluto's status. In his article Vergano quoted planetary scientist Larry Esposito of the University of Colorado, who did not mince his words: 'If Pluto were discovered today, it would not be classified as a planet.' At the press conference, Stern tried a counteroffensive ('It's round. It has a satellite. It has an atmosphere. Thus, it is a planet.'), but the damage was already done. That same evening, the CBS news devoted more attention to Pluto's possible sad fate than to Stern and Buie's spectacular observations.

The Pluto question continued to simmer until early 1998. At the Minor Planet Center, Brian Marsden had not given up hope that Pluto could one day be added to the list of asteroids and other small fry in orbit around the Sun. For him, that was where it belonged, just like all the other ice dwarfs beyond the orbit of Neptune. After all, it would be pretty strange if all known Kuiper Belt objects were to be included in the catalog except the largest one of all.

Marsden did not of course want to create the impression that he had something personal against Pluto. On the contrary, the first ice dwarf to be discovered should have a place of honor in the catalog. By 1998, more than 9,000 asteroids had been given official numbers, and those with round numbers had also been given special names like Piazzia and Herschel (after the discoverers of Ceres and Uranus), Leonardo, Isaac Newton, and IAU. Marsden thought number 10000 should be reserved for Pluto.

He discussed his proposal with Mike A'Hearn of the University of Maryland in Greenbelt, who was then president of Division III, the IAU department concerned with planetary systems sciences. A'Hearn was immediately enthusiastic and did not expect a great deal of opposition to Marsden's idea, especially if Pluto were to be given a kind of 'dual classification.' Why could it not be included in Marsden's list of small solar system object as an ice dwarf and still retain its planetary status?

Pluto Soap Opera

It would not be the first time that a celestial body would be included in two catalogs. The large centaur Chiron, discovered by Charles Kowal in 1977, is known as asteroid number 2060 but, because it had appeared to act like a comet at the end of the 1980s, is also referred to as comet 95P/Chiron. Asteroid 1979 VA, discovered at Lowell Observatory, also appeared to have been recorded earlier, in 1949, as comet Wilson-Harrington. Comet Elst-Pizarro is the same object as asteroid 1979 OW_7. And, although asteroid 1996 PW has no cometary designation, it moves in an elongated cometary orbit around the Sun. Surely there were sufficient precedents to include a bona fide planet in the catalog of small bodies?

But somehow the idea leaked out in the fall of 1998 and the Plutophiles got wind of it before it could be put down on paper and supported with well-reasoned arguments. Colleagues were mobilized to oppose the 'disastrous' plan and the story was broadcast that Marsden was set on demoting Pluto.

A'Hearn made several valiant attempts to discuss the arguments for and against with some of the more prominent opponents, but to no avail. Emotions ruled the day, the discussion became overheated and the atmosphere increasingly hostile. Letters and e-mails from schoolchildren started to pour in: who dared to depose the favorite planet of America's youngsters?

Mike A'Hearn (Photo archive of the Division for Planetary Sciences, AAS)

An e-mail survey of all the members of the IAU at the end of 1998, put together in a hurry by Mike A'Hearn, left no room for doubt: the majority of the respondents – most of whom were Americans – were not in favor of giving Pluto an 'asteroid number.' With a decision still not taken and the numbered asteroids rapidly approaching the 10,000 mark, confusion only increased. Not only among astronomers, but also among the general public, who were kept up to date on the latest developments in the Pluto soap opera by the media.

On Monday, February 1, 1999 (10 days before Pluto would be further away from the Sun than Neptune for the first time in 20 years) the Division for Planetary Sciences (DPS) of the American Astronomical Society decided that enough was enough. In an official statement to the IAU, the society's board disassociated itself from Marsden's proposal and DPS chairman Don Yeomans of NASA's Jet Propulsion Laboratory called on its members to vote against it on A'Hearn's specially set-up webpage.

A day later, Marsden made a final attempt to get his way, in the form of an 'editorial note' to the knowledgeable subscribers of the *Minor Planet Circulars*, but it was all to no avail. On Wednesday, February 3, the IAU issued a press release in which general secretary Johannes Andersen diplomatically made it clear that no division, working group or committee at the IAU was prepared to make any changes whatsoever to Pluto's status. The ninth planet would definitely not be allocated an asteroid number. Pluto's hide appeared to be saved. At least for the time being.

By now, it was becoming increasingly clear that the concept of 'planet' had never been clearly defined. Is it enough for a body to be orbiting the Sun? Of course not, otherwise all asteroids, comets, ice dwarfs, and interplanetary dust particles would also be planets. Does a planet have to have an atmosphere? No, because then Mercury would not count either. Does it have to have one or more satellites? Obviously not, or Venus would have to join Mercury on the bench. Or was it necessary to be of a certain minimal size to be considered a planet? If so, where was the limit? At 2,000 kilometers, so that Pluto would still count? Isn't that a bit of a random choice?

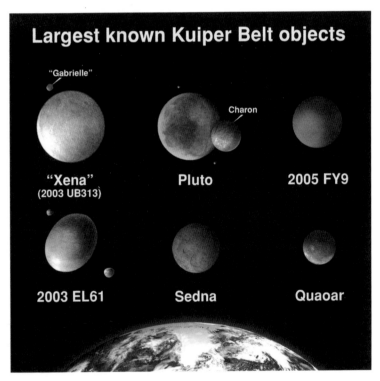

Size comparison of the largest Kuiper Belt objects. The Earth is drawn to scale (NASA, ESA, and A. Field (STScI))

Furthermore, shouldn't the definition depend on an object's orbit? Jupiter's moon Ganymede and Saturn's moon Titan are both larger than Mercury and, geologically speaking, a lot more interesting. In addition, Titan has a thick atmosphere, while there may be something resembling continental shift on Ganymede. If these two satellites orbited the Sun, they would certainly be classified as planets. Obviously it's not as easy as it might seem to decide when a planet really is a planet and dynamical considerations also play an important role.

The discussion flared up again in August 2000, at the 24th general meeting of the IAU in Manchester, England. Mike A'Hearn, who was stepping down as President of Division III, gave a lecture in which he once again argued for Pluto to be given a dual classification. Like all the other presentations at the three-yearly conference, his lecture – entitled 'Pluto: a planet or a *trans*-Neptunian object?' – would be published in the 12th part of the IAU series *Highlights of Astronomy*. Alan Stern, of course, found that unacceptable and he called for a counter-argument to be published in the same volume. The IAU agreed and in September Stern wrote an article, together with colleague Hal Levison, proposing an objective definition of the concept of 'planet.' It was, of course, a definition which included Pluto.

In simple terms, Stern and Levison proposed that a planet must orbit the Sun and be large enough that its shape was primarily determined by gravity. In other words, the object must have adopted a more or less spherical shape as a consequence of its own gravity. In practice that boils down to a minimum diameter of 700–800 kilometers, at least for a body consisting mainly of rock. In addition, the two astronomers made a distinction between the eight 'solitary' planets (from Mercury to Neptune) and those that share their orbital regions with a large number of smaller objects of the same type, such as Pluto and the large spherical asteroid Ceres.

It was perhaps a little unfortunate that Stern and Levison (who both have Jewish origins) proposed the names '*über*-planet' and '*unter*-planet' for the two categories. But their proposed classification came to grief in any case in a confusing mishmash of prefixes and qualifications. Their article spoke of planetary bodies, unbound planets, *über*-giants, and *unter*-subdwarfs. It certainly didn't make the whole issue a great deal clearer.

Wanted: A Definition

It was Neil deGrasse Tyson, director of the prestigious Hayden Planetarium in New York, who finally put the cat among the pigeons. He was responsible for a brand-new exhibition on the solar system at the Rose Center for Earth and Space, part of the renowned American Museum of Natural History and home to the Planetarium. The exhibition made no bones about it, informing hundreds of thousands of visitors that the solar system had eight planets and that Pluto was just one of the countless ice dwarfs beyond the orbit of Neptune. No more nonsense about dual classification – it was time the general public were told the truth, the whole truth, and nothing but the truth.

Brian Marsden was of course pleased as punch with Tyson's decision, but when *The New York Times* devoted considerable attention to the popular museum's controversial position on January 22, 2001, it unleashed a wave of critical reaction. Astronomer and publicist David Levy thought that Tyson was so far wide of the mark it was 'like he's in a different universe.' This was of course no surprise: Levy had written a biography of Clyde Tombaugh and, just before the old astronomer died at the age of 90 in January 1997, Levy had promised him that he would fight tooth and nail to oppose the demotion of Pluto. Campaigns were set up to gather signatures and petitions were submitted, and the debate once again escalated, dominated by cultural, historical, educational, and, above all, emotional arguments. But the exhibition at the Hayden Planetarium went ahead unchanged.

And what about the long-awaited, all-encompassing definition of 'planet'? In the years that followed, it still failed to materialize. Countless professional astronomers and well-intentioned amateurs tried their hand at it, but all to no avail. It was starting to look like a mission impossible. Meanwhile, the need for a clear definition was becoming increasingly urgent. At the end of 2000, the large ice dwarf Varuna was found, followed in the spring of 2001 by Ixion, and on June 5, 2002 and July 26, 2003 the first images of Quaoar and Orcus were recorded on Palomar Mountain. Every one of these objects was larger than 700 kilometers in diameter, and therefore qualified as *unter*-planets in the terminology proposed by Stern and Levison. It was clear that before long a Kuiper Belt object would be discovered that was larger than Pluto. Before that happened, it was essential to establish once and for all when a celestial object is a planet and when not.

When, on March 15, 2004, Mike Brown and his colleagues announced the discovery of Sedna (which potentially had a diameter of some 1,800 kilometers), the IAU finally took action. The board of Division III set up a committee to address the issue of when a planet really is a planet. The committee was chaired by the British astronomer Iwan Williams, who had discovered the fifth and sixth ice dwarfs in 1993 and was then the president of Division III. Among the 18 other members of the committee were comet expert Mike A'Hearn, infrared astronomer Dale Cruikshank, theoretician Alan Boss, Lowell director Ted Bowell, Plutophiles Alan Stern and – how could he have been left out? – Brian 'Pluto-*basher*' Marsden.

Nineteen intelligent, eloquent, and opinionated people with 19 different opinions who never actually spoke to each other in person, but communicated by email – it was a recipe for disaster. Williams could not even manage to organize the numerous voting rounds to everyone's satisfaction (Stern still gets agitated when he recalls the disorderly way in which the committee was led) and his efforts to get all the committee members to agree were a complete failure. And to top it all, in the summer of 2005, after more than a year of interminable discussion, the discovery of Xena (2003 UB_{313}) was announced. Xena was larger than Pluto, status provisionally unknown.

In November 2005, Williams' committee published its final report. It offered no unanimous recommendation, only a summary of the various viewpoints and alternative proposals for classifying planets. In effect, Williams put the ball back in the court of the IAU's executive board. The report was given little publicity and the committee quietly disbanded. Marsden could do nothing about the fact that 2003 UB_{313} could still not be given an official name. No one knew what to do about Pluto and Xena, and the IAU board was completely at a loss.

It was time for last-resort measures and a final deadline. If 19 experts could not reach a decision, and if cultural and historical considerations were so important, perhaps it was better to set up a much smaller working group that did not consist solely of astronomers. So Pluto's fate was placed in the hands of a 'committee of wise men,' who would consult in the deepest secrecy. The seven members were given very clear instructions to come up with a concrete proposal for a definition of what constitutes a planet by July 2006. A final decision would be taken in August at the 26th annual meeting of the International Astronomical Union in Prague. It was D-Day for Pluto and Xena.

It sounded perfect. Too perfect perhaps. But that the committee's final proposal would be rejected right at the last minute – and that the number of planets in the solar system would increase from 9 to 12 and then fall again to 8 in the period of 1 week – was something no one could have predicted. In Prague, the war of the worlds erupted in its full fury.

Chapter 27
Planetary Elections

Owen Gingerich wasn't in Prague on August 24, 2006. He followed the bizarre 'planetary elections' at the General Assembly of the International Astronomical Union (IAU) via the IAU's webcast. And it was pretty depressing. Right up to the last minute, Gingerich – Professor Emeritus of Astronomy and of the History of Science at Harvard University – had argued for Pluto to retain its planetary status, but it had all been in vain. That Thursday afternoon, at 3.32 pm European time, the curtain finally fell for the smallest and most remote planet in the solar system. Pluto was no longer one of the family.

Owen Gingerich (Harvard-Smithsonian Center for Astrophysics)

And it could all have turned out so much better. Under the chairmanship of Gingerich, seven brilliant minds had discussed the issue long and hard. None of

G. Schilling, *The Hunt for Planet X*, DOI 10.1007/978-0-387-77805-1_27,
© Springer Science+Business Media, LLC 2009

them had expected their final proposal for the definition of a planet to meet with any serious objections. They had thought of everything: cultural considerations, historical arguments, educational aspects, scientific integrity, you name it. But they had apparently overlooked one thing: planetary dynamics. Or perhaps common sense.

The IAU Planet Definition Committee. *Back row, left to right*: André Brahic, Iwan Williams, Jun-ichi Watanabe, Richard Binzel. *Front row, left to right*: Catherine Cesarsky, Dava Sobel, Owen Gingerich (International Astronomical Union)

They were a mixed bunch, the committee put together in early 2006 by the executive board of the IAU to seek a final definition for what constitutes a planet. Iwan Williams was of course a member because he had chaired the IAU working group that had earlier failed to produce a unanimous recommendation. Then there was Catherine Cesarsky, director-general of the European Southern Observatory and president-elect of the IAU. And the eminent Gingerich, who knew more than anyone about the history of astronomy. French planetary scientist André Brahic and Japanese solar system expert Jun-ichi Watanabe were on the team, as was asteroid expert Richard Binzel of the Massachusetts Institute of Technology. And, surprisingly enough, science journalist and novelist Dava Sobel, whose bestsellers *Longitude*, *Galileo's Daughter*, and *The Planets* had done more for the discipline's PR than many a professional astronomer. All of them renowned names who had won their spurs not only in astronomy but also in science communication.

But even for brilliant minds it is no small task to find – after the fact – a conclusive definition for a concept that has become a part of everyday usage and popular culture. That became clear on Friday, June 30, 2006, when the committee met in deepest secrecy in a room at the stately Paris Observatory, where mathematician Urbain Le Verrier, the man who predicted the discovery of Neptune, had been in charge a century and a half previously. Opinions were divided, at times deeply, the discussion became very heated, and at the end of the day it did not look as though they were ever going to reach a consensus.

But the day after, they succeeded. The last objections were removed, differences of opinion were ironed out, and desperation made way for relief. Gingerich's planet definition committee drew up a draft resolution which was received enthusiastically by the IAU board. They decided that the IAU's members should vote on the resolution at the 26th General Assembly at the end of August. The content of the resolution was, however, kept secret for the time being – no one wanted another heated Pluto debate erupting in the media. A press release was not issued until Wednesday, August 16, the day after the opening ceremony of the General Assembly in Prague.

There was one thing, however, that no one had to concern themselves about: media attention for the IAU proposal. Newspapers and television news broadcasts throughout the world opened with the story. And no wonder – fears that Pluto was to be thrown out of the planet family had proved unfounded. In the new 'twelve-planet' proposal, the solar system did not lose one planet, it gained three. And before long, that number would almost certainly rise from 12 to 20, then to 50, and perhaps in time even to a few hundred.

Alan Stern had said it often enough: ask children to draw a planet and they will start with a circle. Planets are round. And they must be large enough for their own gravity to have formed them into a spherical shape. Or, in scientific terms, into a state of hydrostatic equilibrium. It then seemed arbitrary to include larger spheres but not smaller ones. That would mean drawing a line somewhere at random. This was also how the planet definition committee looked at it. Let Mother Nature decide who was in and who was out. The spherical shape had to be the deciding factor.

Dwarf Planets

But was it not strange that the solar system clearly had eight 'large' planets and a handful of mini-planets, like the asteroid Ceres and the ice dwarfs Pluto and Xena? That may be so, but Gingerich had found an answer to that: Mercury, Venus, the Earth, Mars, Jupiter, Saturn, Uranus, and Neptune could be referred to as the eight 'classical planets,' while Ceres, Pluto, and Xena were 'dwarf planets.' And to emphasize the special status of Pluto and objects like it – after all you had to take account of the sensitive souls of countless Plutophiles – icy dwarf planets with orbital periods of more than 200 years could be referred to as 'plutons,' a new category of celestial body, with Pluto as the prototype.

But wait: eight classical planets, one dwarf planet in the asteroid belt, and two plutons beyond the orbit of Neptune... that comes to 11. So where is number 12? That was probably the biggest surprise in the new proposal: the 12th planet was Charon, Pluto's moon. Of course, a planet must orbit the Sun, so spherical bodies that circle a planet, such as our own Moon or the large moons of the giant planets, do not count. But Charon is enormous in relation to Pluto. So large that their common center of gravity is not deep below Pluto's surface but somewhere halfway between them. In that case, the planet definition committee decided in its infinite wisdom that the two bodies formed a binary planet. And if Pluto and Charon were a binary planet, it was logical that Charon itself must be a planet.

Eight classical planets and four dwarf planets, three of which were plutons and one was a double planet: 12 planets in total and another 12 on the waiting list – it was a controversial proposal to say the least. Those on the waiting list were bodies about which there was not yet absolute certainty that they were in a state of hydrostatic equilibrium, but once it became clear that this was the case, objects like the asteroids Pallas and Vesta, and the ice dwarfs Sedna, Orcus, and Quaoar would of course also be classified as planets. Ice dwarfs are probably spherical once they have a diameter of 400 kilometers. On this basis, according to Mike Brown's calculations, it would then not be long before the number of planets in the solar system would rise to 53. In the long term, as many as a few hundred spherical objects could be found orbiting the Sun.

Brown himself had proposed that all objects larger than Pluto should be called planets, but he could live with the IAU committee's twelve-planet proposal, even though it did make the concept of 'planet' susceptible to inflation. He would in any case go down in history as the discoverer of the first planet beyond the orbit of Pluto. Stern was over the moon with the idea: 'his' space probe New Horizons, which had recently been launched and was now on its way to Pluto and Charon, would now be visiting two planets. From Flagstaff Jim Christy – the man who found Charon – was astounded to hear the news that he now belonged on the illustrious list of planetary discoverers. And Sicilian astronomers must have been pleasantly surprised that Ceres, discovered in 1801 by Giuseppe Piazzi and named after the island's patron saint, would also once again be a planet, just as in the first half of the nineteenth century.

But by no means everyone was happy with the IAU's twelve-planet resolution. Many planetary scientists saw the proposal as an artificial construction with one dubious objective – to keep Pluto on board. It was nonsense to look only at a body's physical properties. The criteria should also include environmental and dynamical factors. It was then crystal clear that there was an enormous difference between the eight planets and the countless asteroids and ice dwarfs in orbit around the Sun. The fact that some of these small bodies happened to be just about large enough to achieve a state of hydrostatic equilibrium did not mean that they suddenly belonged in a completely different category.

Rebellion

On Friday, August 18, the seeds of a rebellion against the twelve-planet proposal were sown at the IAU conference in Prague. During a tumultuous meeting of Division III, the Uruguayan astronomers Gonzalo Tancredi and Julio Fernández presented a counter-proposal signed by nearly 20 colleagues. It described a planet as 'by far the largest object in its local population.' The wording may have been a little awkward, but the idea was clear enough: the eight 'classical' planets are not part of a larger family that orbit the Sun in the same region; because of their much stronger gravity, they are lord and master of their orbital neighborhood. Small spherical bodies which did not meet these criteria, such as Ceres and Pluto, could be called dwarf planets but, at least according to Tancredi and Fernández, the solar system had only eight *real* planets.

That same afternoon, the planet definition committee discussed the objections. They were not restricted to the actual definition, but also the surprising classification of Charon and the use of the term 'pluton.' The latter was considered especially confusing, since Pluto is called 'Pluton' in France and because the term 'pluton' was already used by geologists to refer to a specific kind of rock formation. In consultation with IAU president Ron Ekers, Gingerich and his colleagues decided to re-write the resolution and divide it into two. In the first part, a planet was defined generally as a spherical object in orbit around a star; the second part made a distinction in our own solar system between the eight 'classical planets,' which are dominant in their local environment, and the 'dwarf planets' which are not. In this way, they hoped to satisfy the objections of the dynamicists, while retaining Pluto's status as a planet.

But Gingerich was not to get off that lightly. An extra lunchtime session was arranged for all IAU members on Tuesday, August 22, which turned into a full-fledged confrontation with the rebels. They demanded that the dynamical criterion be given a central role in the definition and therefore be included in the first part of the modified resolution. With of course the logical consequence that dwarf planets, which do not meet this criterion, are not real planets. Italian dynamicist Andrea Milani from the University of Padua didn't quite resort to fist-fighting with Ekers and Gingerich, but it was a close thing.

Under so much pressure from its own members, the IAU had to give in. The resolution was once again re-written, only 2 days before the final vote on Thursday the 24th. The new text made a distinction between 'planets' – spherical bodies orbiting the Sun which were the largest object in their local population zone – and 'dwarf planets,' for which the first two criteria applied, but not the third. During the second session of the extra meeting, at the end of the afternoon, the majority of the IAU members present were willing to accept this definition. Pluto's fate was sealed. Owen Gingerich hardly slept a wink that night.

The following day, like hundreds of other astronomers, Gingerich was his way back to the airport in Prague. The IAU conference lasted 2 weeks and almost none of the more than 2,400 participants stayed for the entire fortnight. Most of the tickets for the closing ceremony on Thursday afternoon had been reserved before it became clear that a vote would be taken on the definition of a planet and the status of Pluto. And no one would ever have expected that emotions would run so high.

Although Gingerich was not present at the conference center in person, he did make a final attempt to secure Pluto's future as a full member of the planetary family. In an e-mail to the IAU resolution committee he made the brilliant suggestion of adding one word to the final text, which would completely reverse the meaning of the resolution.

Foul Play

As a result, on the Thursday morning, the ninth edition of *Dissertatio cum Nuncio Sidereo III* (the official conference newspaper that was distributed to the IAU members each morning), informed its readers that there were now four resolutions relating to the definition of a planet and the status of Pluto, and that there would be a vote on each of the resolutions that afternoon. Resolution 5A was practically identical to what had been put down on paper on Tuesday afternoon: planets are spherical objects in orbit around the Sun which have 'cleared the neighborhood around their orbits'; dwarf planets are in hydrostatic equilibrium, but are not dynamically dominant. In Resolution 5B, however, which was added at the request of Gingerich, it was proposed that the word 'planets' in Resolution 5A be replaced by 'classical planets.' If that was approved, the definition would no longer make a distinction between planets and dwarf planets, but between two sub-classes of classical planets and dwarf planets.

It seemed like an unimportant addition, but the consequences were far-reaching. The results of the voting on Resolution 5B would answer the question whether Pluto is a planet or not. No one put that into words better than Rick Fienberg, editor-in-chief of the monthly magazine *Sky & Telescope*. If 5B were rejected, he said, the answer to the question would be: 'No, Pluto is a dwarf planet.' If 5B were to be adopted, the answer would be: 'Yes, Pluto is a dwarf planet.' And then there would also be a vote on two resolutions which focused fully on Pluto. Resolution 6A described the dwarf planet Pluto as the prototype of a new category of *trans*-Neptunian object, while Resolution 6B gave this category the name 'plutonian objects.'

The backrooms and corridors of a major scientific conference were rarely as rowdy as on that Thursday morning in Prague. The addition of Resolution 5B was seen by its opponents as an attempt to hijack the vote; Julio Fernández even went as far as to denounce it as foul play: if Resolution 5B were adopted, it would have an impact on the content of Resolution 5A, making it impossible to vote on 5A first. At the same time, Fernández and his supporters (including Gonzalo Tancredi and, of course, Brian Marsden) knew only too well how difficult it is to explain to the general public that a 'dwarf planet' is not a planet – a raincoat is, after all, still a coat, and a kitchen chair still definitely a chair. An attempt to replace the term 'dwarf planet' at the last minute by a completely different word, such as 'planetino,' however, came to nothing.

During the closing ceremony of the General Assembly that afternoon, you could cut the tension in the large conference hall with a knife. Votings on the first four resolutions – on a number of technical matters relating to astronomical coordination systems – were just formalities, but then it was time for the 'planet resolutions.' The television cameras were rolling, and journalists sat with their notebooks and mobile telephones at the ready. Within an hour, the whole world would know just how many planets there were in the solar system.

Jocelyn Bell Burnell, cradling a toy Pluto during the closing ceremony of the IAU 2006 General Assembly (International Astronomical Union)

The IAU board was fortunately prepared for the overwhelming media attention. British radio astronomer Jocelyn Bell Burnell, a member of the resolution committee who was leading this part of the session, had done her homework. Telephoto lenses zoomed in on the objects she had laid out on the table: a blue party balloon, a packet of Jordan's Country Crisp cereals, a cuddly toy dog, and an open umbrella. The balloon represented the eight classical planets. The packet of cereals stood for the asteroids (the word cereals is derived from Ceres, the goddess of the harvest). The dog was of course Mickey Mouse's pet Pluto, representing the ice dwarfs, and Bell Burnell was intending to use the umbrella – which had a sign on it saying PLANETS in large letters – to show which objects would fall under this umbrella term.

It was a photogenic presentation, but unfortunately it played for a half empty hall. The large majority of IAU members were already on their way home; Pluto's fate was in the hands of the little more than 400 astronomers who were still in Prague. At the start of the session they had been given yellow cards to hold up in the air to cast a vote. Astronomy students had been called in to count the votes if there were any doubt about whether there was a majority or not.

During the vote on Resolution 5A there was certainly no need for a count; it was adopted with a clear majority. Astronomers now finally had a definition of a planet: a celestial body which is in orbit around the Sun, has sufficient mass to assume a hydrostatic equilibrium, and has cleared the neighborhood around its orbit. Spherical objects that do not meet the third criterion would now be known as dwarf planets; all other asteroids, ice dwarfs, and comets would be classified as 'small solar system objects.'

Then it was time to vote on Resolution 5B. To increase the suspense even more, a supporter and an opponent of the resolution were each given 2 minutes to defend their positions. Richard Binzel, representing the planet definition committee in the absence of Owen Gingerich, made an impassioned plea for inclusion of the word 'classical' in the freshly adopted definition. That would put the packet of cereals and Disney dog in the umbrella after all. But Mark Bailey of the Armagh Observatory in Northern Ireland appealed to those present to vote against the resolution, which he called 'misleading.'

But Bailey's appeal was no longer necessary. When IAU president Ron Ekers put the controversial resolution to the vote, it was rejected by an overwhelming majority. At 3.32 pm on Thursday, August 24, 2006, Pluto's life as a planet came to an end.

There was one small consolation for the tiny object: because Resolution 6A had been adopted, it would be considered the prototype of a new category of *trans*-Neptunian ice dwarfs in hydrostatic equilibrium; in other words, dwarf planets with an orbital period of more than 200 years. But they were unable to find agreement on a name for that new class of objects in Prague. (In 2008, the IAU decided to call these objects 'plutoids'). Resolution 6B had no clear majority – and no one felt the need for a recount or second round of voting.

IAU members voting against Pluto's planetary status (International Astronomical Union)

Perhaps those astronomers who were still present at the conference had had enough of the whole business. Cosmologists and astrophysicists, who concern themselves with problems of a completely different nature, understood little of the emotions these issues unleashed among their planetary colleagues. And even planetary scientists themselves heaved a sigh of relief now that the whole thing had been settled. The Pluto war was over. Over and out. Tomorrow they could get back to work.

Chapter 28
The Continuing Story

Patsy Tombaugh was delighted when, just over 6 months after the tumultuous planetary elections in Prague, Pluto was promoted again – at least locally. The 94-year-old widow of Pluto-discoverer Clyde Tombaugh was in the public gallery with her daughter Annette in Santa Fe when *House Joint Memorial 54* was unanimously adopted by the 70 members of the House of Representatives of the State of New Mexico. It was March 13, 2007, 77 years after Pluto's discovery had been announced by Lowell Observatory. According to the memorial, submitted by Democratic representative Joni Marie Gutierrez from Mesilla, the State of New Mexico (where Clyde and Patsy had lived most of their lives) fully recognized Pluto's planetary status. And it had officially designated March 13, 2007 as 'Planet Pluto Day.'

Patsy Tombaugh (Richard Shope, NASA/JPL, courtesy Max Mutchler)

G. Schilling, *The Hunt for Planet X*, DOI 10.1007/978-0-387-77805-1_28,
© Springer Science+Business Media, LLC 2009

That was good news for Patsy of course, who had very bravely told journalists at the end of August 2006 that she was 'disappointed, but not surprised' at the decision by the International Astronomical Union (IAU). And it was good news for the makers of the websites www.plutoisaplanet.com and www.pleasesavepluto.org, which described the IAU's decision as 'fraudulent' and appealed to everyone visiting the site to sign a petition calling for Pluto's planetary status to be restored. And for director Mark Sykes of the Planetary Science Institute in Arizona and Plutophile Alan Stern of the Southwest Research Institute in Boulder, Colorado, who had circulated a similar petition among their fellow astronomers and obtained hundreds of declarations of support.

Stern had been furious at the way things had gone at the IAU conference in Prague, which he had been unable to attend personally. The original 12-planet proposal may have been drafted without consulting the wider astronomical community, but it was at least well-thought-through, carefully worded, and above-all, favorable to Pluto. But now, in place of that proposal, they were all stuck with a hastily thrown together and badly formulated definition, which made no clear distinction between planets and dwarf planets, and which had been adopted on the basis of the votes of only 424 members of the IAU.

When Stern was asked to re-write his contribution to an American textbook on astronomy in light of the new definition, he withdrew his collaboration out of protest. And like many of his colleagues, he was not intending to use the makeshift definition himself. The petition presented to the IAU board by fellow Pluto supporter Mark Sykes, which had been signed by more than 300 astronomers in only 5 days, made it abundantly clear that a better definition was required. Preferably a definition that was reached in an atmosphere of complete openness and with feedback from anyone with an interest in the issue. It could then be put to the vote at the following IAU General Assembly in Rio de Janeiro in the summer of 2009.

Eris and Dysnomia

To many people's surprise, Mike Brown was not among those who signed the petition. Of course it was a shame that 2003 UB_{313} could no longer claim to be the tenth planet (Brown published an emotional 'Requiem for Xena' on his website the day after the IAU conference), but he was perfectly happy with the new definition, even though the wording might leave a little to be desired. An error made 76 years previously had finally been put to rights. Astronomers had made the correct decision, even though it might be a little painful. And of course Xena had by no means disappeared from the scene; as the largest of the dwarf planets, it continued to be an extremely interesting object.

The name, of course, had to go. Now the status of the icy body had been determined, it should no longer be referred to with such a light-hearted nickname. Bad luck for the New Zealand actress Lucy Lawless, who had played the title role in the cult series *Xena: Warrior Princess*. She had congratulated Brown and his colleagues personally on their spectacular discovery and was naturally honored by their choice of a temporary name. But fortunately Brown came up with a perfect name to replace Xena. Two weeks after the IAU conference, he proposed the name Eris, after the Greek goddess of discord and strife. A more appropriate name was hardly conceivable for an object that had been the cause of the bitterest dispute in the history of planetary science.

Artist impression of Eris and Dysnomia (NASA, ESA, and Adolf Schaller)

Xena's satellite, which had been called Gabrielle after the warrior princess's sidekick, also had to be given another name of course. It was rechristened Dysnomia, the daughter of Eris and the demon of lawlessness. Again, a very suitable mythological name – and a tongue-in-cheek reference to the much-admired Lucy Lawless. And Brown had not forgotten that Jim Christy had named Pluto's moon after his own wife. If you shortened Dysnomia to 'Dy,' it sounded the same as 'Di,' his wife Diane's nickname.

On Wednesday, September 13, 2006, both names were officially announced and Eris was given the minor planet number 136199. A week previously, without a great deal of publicity, the second largest dwarf planet – Pluto – was also given its own number: 134340. No nice round numbers like those allocated to Varuna and Quaoar, but 'normal' ones that were difficult to remember, as if Brian Marsden and his colleagues wanted to make it extra clear that dwarf planets also had to take their turn in the long list of solar system objects. Pluto now inhabits an inconspicuous position between two small asteroids: 134339, discovered on Palomar Mountain in 1977, and 134341, found 2 years later by the Siding Spring Observatory in Australia. There's nothing a memorial from the New Mexico House of Representatives can do to change that.

Now (the spring of 2008) 132 ice dwarfs have been given an official number, and that is certain to increase in the future as the orbits of more objects in the Kuiper Belt are determined with sufficient accuracy. Of course, the number of ice dwarfs without an official designation is rising even more rapidly. By March 2008 almost 1,100 *trans*-Neptunian objects had been identified, and every few weeks, when the night sky around the New Moon is dark enough to hunt for extremely faint moving dots of light, even more are found. As new searches are instigated, the number of ice dwarfs is expected to have risen to tens of thousands within a few years; plenty to conduct statistical research.

This has not been very easy up till now because of the search strategies applied. Brown and his colleagues used the Schmidt Telescope on Palomar Mountain to survey the entire night sky, but they were only looking for the brightest ice dwarfs. Other surveys may be more sensitive to much smaller, fainter objects, but have to focus on a more restricted part of the sky. To track down faint ice dwarfs using current technology, you need a small field of view and a relatively long exposure time, which makes it impossible to cover the entire night sky. On the other hand, there are far more small ice dwarfs than large ones.

Buffy

A good example of such a 'deep' survey was the Canada–France Ecliptic Plane Survey, conducted with the 3.6-meter Canada–France–Hawaii Telescope on Mauna Kea. In the summer of 2005, when the first phase of the survey was completed, the telescope had scanned a total of 410 square degrees – approximately 2,000 times the surface of the Full Moon. That may sound like a lot, but it is only 1% of the entire night sky. Under the leadership of Brett Gladman, by then associated with the University of British Columbia in Vancouver, the Canadian–French survey unearthed a few hundred faint ice dwarfs. A follow-up project with a much larger electronic camera should find many thousands more.

At the end of 2004, the Canadian–French team came across a relatively bright ice dwarf on images recorded on December 11 and 12. Gladman's colleague Lynne Jones, who discovered the object when analyzing the images, saw immediately that it was exceptional. 2004 XR_{190}, as the object was called provisionally, was not only brighter than most of the other ice dwarfs found in the survey, but it also appeared to be at a considerable distance from the Sun: more than 8.8 billion kilometers (59 astronomical units (AUs)). The combination of a great distance and relative brightness suggested that it had a diameter of as much as 1,000 kilometers.

On top of that, 2004 XR_{190} proved to circle the Sun in an extraordinary orbit. Follow-up observations in 2005 showed that it was almost circular. Even at its closest point to the Sun, the ice dwarf is still more than 7.8 billion kilometers (52 AU) away – far beyond the sphere of influence of Neptune. Although the orbit was not eccentric, however, it was very highly inclined, at an angle of almost 47 degrees. This was clearly a remarkable discovery. Inspired by Mike

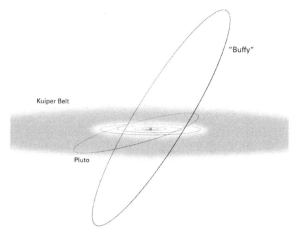

The unusual orbit of 2004 XR$_{190}$ ('Buffy') (Wil Tirion)

Brown, who had come up with the name Xena in the summer of 2005, Jones and Gladman nicknamed their ice dwarf Buffy, after the main character in the television series *Buffy: the Vampire Slayer*.

The discovery of Buffy was announced on Tuesday, December 13, 2005. Gladman and his colleagues had no satisfactory explanation for the strange orbit. If Buffy's motion could not now be affected by Neptune, it could not have been forced into its current orbit by the giant planet either. Perhaps the orbital aberrations were caused by a passing star, but then you would expect to find a lot more Kuiper Belt objects in such peculiar orbits.

That was actually not so improbable as it may seem, according to the team's paper in *Astrophysical Journal Letters*. The Canada–France Ecliptic Plane Survey focused exclusively on the parts of the night sky close to either side of the ecliptic – the orbital plane in which most planets and ice dwarfs rotate around the Sun. In its extremely inclined orbit, Buffy only passes through this zone twice per rotation. For the other 98% of the time, it is far to the north or south of the ecliptic. If Buffy were the only ice dwarf in such a steep orbit, its discovery would have to be an incredible stroke of luck. It seemed more likely that there are countless Buffy's orbiting the Sun.

To find out just how many ice dwarfs are moving in steep orbits, you need to scan the entire night sky for almost invisible slow-moving points of light. That will finally provide a good picture of the structure of the Kuiper Belt, of the different ice dwarf families and their orbital properties, and of the range of sizes of all these *trans*-Neptunian objects. It is like the people of the United States: if you want to know them really well, you should not just concern yourself with the ups and downs of a handful of celebrities, nor is it enough to interview all the participants in, for example, the New York marathon. What you need to do is conduct a population survey that is as complete and representative as possible.

A Hundred Thousand Ice Dwarfs

Such a 'census' of the Kuiper Belt is to be carried out in the next few years by the
University of Hawaii's Panoramic Survey Telescope & Rapid Response System
(Pan-STARRS) project. When it is fully functional, the system will consist of
four identical telescopes, each with a mirror 1.8 meters across. The focal planes of
the four telescopes will be fitted with gigantic electronic cameras with no less than

The Pan-STARRS prototype telescope on Haleakala (Brett Simison/University of Hawaii)

1.4 billion pixels. They will be able to photograph an area of the night sky 15 times the size of the Full Moon in less than a minute. The images will show objects 15 million times fainter than the faintest stars that can be seen on a clear night with the naked eye. Every month, the night sky above Hawaii will be scanned three times, after which supercomputers will scour the many terabytes of observing data for moving points of light.

Pan-STARRS focuses mainly on the search for Earth-grazers: small asteroids that cross the Earth's orbit. They move much faster across the sky than the distant ice dwarfs in the outer regions of the solar system. But, with a slightly adapted observing strategy and simple adjustments to the analysis software, it must be possible to track down at least 20,000 ice dwarfs. A simple prototype of the Pan-STARRS telescope has already been put into operation on the summit of the Haleakala volcano on the island of Maui. It is the intention that Pan-STARRS 4 – the final version with four telescopes – will be located on Mauna Kea on neighboring Big Island. The observatory will replace the old 2.2-meter telescope operated by the University of Hawaii, with which Dave Jewitt and Jane Luu discovered QB_1 long ago, in 1992. They did that with a camera with only 4 million pixels – 350 times fewer than the Pan-STARRS cameras.

Artist impression of the Discovery Channel Telescope (Lowell Observatory)

The ice-dwarf hunters at Lowell Observatory also have big plans. With their Deep Ecliptic Survey, for which they use the 4-meter telescopes on Kitt Peak in Arizona and Cerro Tololo in Chile, they have discovered many hundreds of ice dwarfs, but soon that number will increase to a few hundred *every week*. Sixty-five kilometers to the southeast of Flagstaff, on a high plateau in the forest near the hamlet of Happy Jack, work is under way on the 4.2-meter Discovery Channel Telescope, which is planned to go into operation in 2010. Like Pan-STARRS, this wide-field telescope, sponsored by Discovery Channel, will also seek out Earth-grazers, dark matter, and planets around other stars.

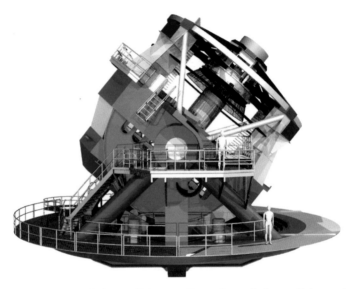

Design of the Large Synoptic Survey Telescope (Large Synoptic Survey Telescope)

And if sufficient funding can be secured, the Large Synoptic Survey Telescope (LSST) will become operational in 2013 on Cerro Pachón in Chile. The LSST will outstrip all of its predecessors: with a mirror 8.4 meters wide, a field of view of 10 square degrees, and a 3 gigapixel camera, it will scan the entire night sky twice a week. No rapid Earth-grazing asteroid, sluggish ice dwarf, or unexpectedly exploding supernova will escape its attention. To analyze all the observing data – a mere 30 terabytes each night – a unique joint venture has been created with the data-magicians at Google.

With the LSST it should be possible to discover at least 100,000 Kuiper Belt objects only tens of kilometers across. Smaller ice dwarfs at billions of kilometers distance from the Sun and the Earth are no longer visible, but resourceful astronomers have found a solution to that problem, too. Very occasionally, one of these frozen ice balls will move in front of a distant star, making the star invisible for a fraction of a second. You can keep track of these stellar occultations by using a series of telescopes to continually observe many thousands of

The Large Synoptic Survey Telescope will be built at Cerro Pachón in Chile (Large Synoptic Survey Telescope)

stars on clear nights. That is the objective of the Taiwan–American Occultation Survey (TAOS) project, which has started at Lulin Observatory in Taiwan. Once a number of teething problems have been solved, the project should provide usable statistical information about the number of small comets in the Kuiper Belt.

So good times are on the way for solar system scientists. Twenty-five years after the discovery of the first ice dwarf (not counting Pluto) the number of known *trans*-Neptunian objects will probably exceed 100,000. By way of comparison, after the discovery of Ceres, it took more than 200 years for the number of known asteroids to increase to more than 100,000. Knowledge of the orbits, dimensions, and surface properties of all these ice dwarfs provides a wealth of information on the evolution of the early solar system, because it is now finally possible to test a wide range of theoretical models in practice.

But the enormous increase in the frequency with which ice dwarfs are being discovered also means that study of the Kuiper Belt will become routine. New discoveries will no longer make the front pages and names like Dave Jewitt and Mike Brown will soon become as much a part of astronomical history as Giuseppe Piazzi, Clyde Tombaugh, and Charles Kowal. The age of the pioneers is over, and with it the excitement.

Unless of course, tens or hundreds of billions of kilometers from the Sun, a world as large as Mars or the Earth is discovered. The hunt for Planet X will probably never come to an end.

Chapter 29
A Thousand Planets

Alan Stern would have preferred to publish his 'Thousand Plutos' article in *Nature* or *Science*. Then at least it would have attracted a little attention, among both his colleagues and the public at large. Which is always a good thing for a young and ambitious postdoc. But Glen Stewart of the University of Colorado in Boulder suggested that it would be better to just submit the article to *Icarus*, the smaller professional journal of solar system studies. After all, the article was not that revolutionary or convincing. And then, *Icarus* editor Joe Burns asked Stern if he was really sure he wanted to publish it.

But Stern did not begin to doubt himself or his ideas. That just wasn't in his character. So the article 'On the Number of Planets in the Solar System: Evidence of a Substantial Population of 1000-kilometer bodies' was published in April 1991 and eventually received quite some attention from the popular-scientific press. And with good reason, in Stern's view. Fifteen years later, the theory is still as solid as a rock: the solar system probably has around a thousand Pluto-like planets. And perhaps a few as large as Mars or the Earth.

Alan Stern has never lacked ambition. In 1970, as a 12-year-old boy, he wrote his first book, *Unmanned Spacecraft – An Inside View*, which was typed out by his grandfather's secretary. His only aim was to be placed on NASA's press list. He was determined to become an astronaut, devoured *Aviation Week & Space Technology*, studied physics and aeronautical and space science at the University of Texas in Austin and, in the summers of 1979 and 1980, worked as a trainee at the Johnson Space Center in Houston, the Mecca of manned space flight.

Okay, becoming an astronaut was perhaps a little over-ambitious (although Alan did get his pilot's license – after all, you never know), but astronomy and planetary science were an excellent alternative. In the mid-1980s, at the age of only 28, Stern was already project scientist for Spartan-Halley, a satellite which was to conduct observations on Halley's Comet. The satellite was lost, however, in the tragic accident with the space shuttle Challenger in January 1986. In December 1989, Alan was awarded his PhD at the University of Colorado for a study of the evolution of comets.

G. Schilling, *The Hunt for Planet X*, DOI 10.1007/978-0-387-77805-1_29,
© Springer Science+Business Media, LLC 2009

Alan Stern (P.E. Alers/NASA)

His article concluding that there must be about a thousand Plutos in the solar system was published in *Icarus* a little over a year later, just before Stern moved to the Southwest Research Institute. And that means a thousand planets, because Alan Stern is not interested in Pluto demotions or vague International Astronomical Union (IAU) definitions. And the fact that all of these ice planets may never be visible from the Earth makes no difference at all. The important thing is that they have left their marks on the outer regions of the solar system.

Indirect Clues

One of these silent witnesses is Pluto's large moon, Charon. Alan was 19 when Jim Christy discovered Charon. Christy was his hero and his passion for Pluto has never waned. But it did not become clear until the 1980s that Pluto and Charon were an exceptionally improbable couple. In 1984, Bill McKinnon showed that the binary planet could only have been created as the result of a collision between Pluto and another large ice dwarf – a catastrophe that probably took place in the early years of the solar system. In his article, Stern calculated that the chances of such a collision were negligible, unless there were already a very large number of large ice dwarfs orbiting the Sun.

A second clue, according to Stern, was the axial tilt of Uranus. Uranus 'lies on its side.' i.e., its axis is almost in its orbital plane, instead of perpendicular to it. This exaggerated tilt is assumed to have been caused by a cosmic collision in a far distant past – and again you can conclude that such an impact was very improbable if there were not a lot of objects flying around in the outer regions of the solar system.

And then, of course, there was the large Neptunian moon Triton, which orbits in the reverse direction. Triton is possibly a large ice dwarf which was captured by Neptune shortly after the planets were created. But no matter how that happened exactly, the chances of capturing such an icy treasure are improbably small, unless there were swarms of Triton-like objects in Neptune's neighborhood.

Three indirect clues, each of which may not be very convincing on their own, but which together strongly suggest that the outer regions of the solar system were once populated by many hundreds, or even thousands of large ice dwarfs a thousand kilometers or more in diameter. So where are all these ice planets now? The answer is simple: just like the majority of the smaller ice dwarfs they were ejected from their orbits by the effects of Neptune's gravity. Some ended up in the inner solar system (perhaps the Sea of Tranquility on our own Moon was created by the impact of one of Stern's 1000 Plutos). Most of them ended up in the Oort Cloud, billions of kilometers from the Sun.

That was before the discovery of 1992 QB_1, when the existence of the Kuiper Belt had not yet been proved. Purely on the basis of logical thinking Stern concluded that, beyond the orbit of Neptune, planets had also aggregated from the solar nebula – the rotating cloud from which the solar system was born. And that, although that process did not result in the formation of a fifth giant planet, nor did it stop at the formation of small, icy planetesimals a few kilometers across. Hundreds of substantial bodies must have been formed, including Pluto and Triton (and Eris!), and that means that there must also have been countless smaller ice dwarfs a few hundred kilometers in diameter.

Since then, Argentine astronomer Adrián Brunini has suggested that Uranus was pushed onto its side by gravitational disturbances from Saturn. According to the Nice model devised by Hal Levison and Alessandro Morbidelli (see Chapter 25) those disturbances were especially strong 700 million years after the planets were formed. The computer simulations that Brunini published in *Nature* on April 27, 2006 offer a very plausible explanation for the axial tilt of the giant planets. That sheds some doubt on Stern's argument on Uranus in the *Icarus* article, but not on his other two 'clues.' On the contrary, the discovery of countless companions and binary objects in the Kuiper Belt suggests that there may indeed have been a large number of collisions in the distant past.

Study of *trans*-Neptunian objects therefore offers an insight into the origins of the solar system. But we by no means yet fully understand all the characteristics of the Kuiper Belt. No one, for example, has ever come up with a satisfactory explanation for the Belt's sharp outer edge. At distances of more than 7.5 billion kilometers (50 astronomical units (AUs)) you suddenly

Triton, the largest moon of Neptune, as imaged by NASA's spacecraft Voyager 2 in 1989
(NASA/JPL/USGS)

encounter no more ice dwarfs, while you would expect the number of objects to
taper off gradually as you go further away from the Sun. That means that the
primal cloud from which the solar system was formed must also have had a
sharp outer edge or that, for some reason or another, accretion never occurred
beyond a certain distance.

The most likely explanation is that the solar nebula was 'kept in check' by the
influence of one or more stars in the immediate neighborhood of the new-born
Sun. The energy-rich radiation of the massive hot star could have blown away
the thin outer parts of the nebula, or the sharp outer edge could have been
created by the gravitational influence of a star passing by at a relatively short
distance. Both ideas are less far-fetched than they may seem at first glance.
Now, the nearest stars are a few light-years away, but that has not always been
the case. Like most other stars, the Sun was almost certainly born in a compact
star cluster. Shortly after it was created, it must therefore have had a large
number of close neighbors – so it is not so absurd to assume that they may have
left their mark in some way or another.

However it happened, the current architecture of the Kuiper Belt is a fossi-
lized reminder of phenomena and processes from the early years of the solar
system. The various families of ice dwarfs, each with their own populations, the
variety of orbital eccentricities and inclinations, the colors and reflective properties of the individual objects, and the variation in scale – they all offer
planetary scientists a glimpse of a distant past. They also impose restrictions
on the scope available to theoreticians who attempt to reconstruct the history of
the solar system. A theory that is unable to reproduce certain characteristics of
the Kuiper Belt can go straight into the garbage can.

Origins

For the most recent insights into the origins of the solar system, we need to go back some 4.5 billion years, to the time when it was formed from a contracting, rotating disk of gas and dust. For some reason or another, the disk had a relatively sharp outer edge at about 4.5 billion kilometers from the Sun (30 AU) – that is, roughly where Neptune's orbit now lies. The solid particles in this solar nebula (metals and silicates in the hot inner regions, ice crystals in the cold outer zones) clumped together to create trillions of planetesimals, the building blocks of the planets.

Close to the Sun, the rocky planetesimals accreted into a large number of protoplanets a few thousand kilometers in diameter. These later merged to form the Earth and the other three terrestrial planets, Mercury, Venus, and Mars. Our Moon was created as the result of one of the last encounters between a protoplanet and the newly formed Earth.

The outer regions of the solar nebula – beyond the 'snow line' – were populated by many more icy planetesimals. Most of them merged to form the cores of Jupiter, Saturn, Uranus, and Neptune. These cores were so large and massive that they attracted enormous quantities of hydrogen and helium gas before the gassy components of the nebula were blown away by the increasingly powerful radiation from the Sun. The giant planets that emerged from this process were closer to each other and to the Sun than they are now: all four were in orbit at a distance of between 1.8 and 3 billion kilometers (12–20 AU).

After the eight planets had been formed, they effectively cleaned up the remaining planetesimals, either by 'swallowing them up' or by catapulting them inward into the Sun or outward, where they became part of the Oort Cloud. Only between the orbits of Mars and Jupiter and beyond the orbit of Neptune were there still broad belts of rocky and icy planetesimals which the planets could not capture. Some of them aggregated to form larger objects, but they did not produce any full-fledged planets.

Over time, as a result of gravitational interaction with the countless clumps of ice in their vicinity, the orbits of the giant planets changed. Jupiter moved a little closer to the Sun, while Saturn, Uranus, and Neptune moved further away. Seven hundred million years after the formation of the planets, Saturn's orbital period was precisely twice that of Jupiter. The corresponding orbital resonances caused chaos in the solar system. For a long time, Uranus and Neptune moved in elongated, inclined orbits before they eventually 'came to rest' further away from the Sun. Enormous numbers of ice fragments from the Kuiper Belt were strewn about the solar system and the asteroid belt was seriously shaken up. The inner solar system was subjected to a cosmic bombardment, the scars of which can still be seen on the Moon.

Artist impression of a collision in the Kuiper Belt (Dan Durda)

As Neptune migrated outward, its gravitational effect caused the remaining ice dwarfs in the Kuiper Belt to be swept along before it. A large number of them ended up in more or less stable orbits around the Sun, at distances of between 4.5 and 7.5 billion kilometers (30–50 AU). Besides the 'classical' Kuiper Belt objects, a large number of ice dwarfs – including the plutinos – were captured in orbital resonance with Neptune. But by far the majority of them ended up in elongated and highly inclined orbits in what is now known as the 'scattered disk.'

So what about the large ice dwarfs? What happened to Alan Stern's 'thousand Plutos'? The same fate befell them as the countless small clumps of ice.

Most of them were catapulted into space at an early stage of the process and now drift around in the distant, cold, and dark Oort Cloud. Some of them, like Eris, have ended up in the scattered disk. Another large ice dwarf – the retrograde Triton – was long ago annexed by Neptune. And yet another – Pluto, together with its giant moon Charon – was captured in a 3:2 resonance with the giant planet.

And of course we must not forget the centaurs. Like the scattered disk objects, they are ice dwarfs expelled from their original orbits by Neptune's gravity. But the centaurs ended up in *smaller* orbits. This sealed their fate: their elongated, elliptical orbits are unstable in the long run. One day they will fall into the Sun or be catapulted into space, at least if they do not collide with a planet first.

Mysteries

It all sounds coherent and watertight. As though planetary scientists have finally cracked the riddle of the origins and evolution of the solar system. But of course it is not as simple as that. There are still many unanswered questions and unsolved mysteries. Research into the trends and patterns in the surface properties of ice dwarfs is still in its infancy and the reasons for the sharp edge of the Kuiper Belt, at 7.5 billion kilometers from the Sun, remain obscure.

And that applies too to the existence of ice dwarfs in extremely distant orbits, such as 2000 CR_{105}, Sedna, and Buffy. Even at the closest point of their orbits, they are still more than 6.5 billion kilometers from the Sun (43 AU), far beyond Neptune's sphere of influence. Sedna never gets closer than 11.4 billion kilometers (76 AU). The scattered disk objects also spend most of their time at great distances from the Sun, but at the nearest points of their orbits they are not as far beyond Neptune. They are therefore regularly within the giant planet's sphere of influence, the strongest indication yet that it was Neptune that forced them into their current orbits.

2000 CR_{105}, Sedna, and Buffy, however, orbit the Sun at such great distances that they never feel the effects of Neptune's gravity. It is therefore inconceivable that they would end up in such extremely remote orbits as the consequence of an encounter with the outermost planet. That begs the question, of course, of what did cause them to adopt their current orbits. And these objects are by no means the exception that proves the rule. 2000 CR_{105}, Sedna, and Buffy are without doubt the minute tip of a colossal iceberg. They are all reasonably large, probably between 400 and 1,800 kilometers in diameter. And, when they were discovered, they were close to the nearest point of their orbits. There must therefore be countless similar objects that are either much further away on their elongated elliptical orbits or considerably smaller, making them far too faint to be detected with current techniques.

The origin of the 'extended scattered disk,' as this population of ice dwarfs is called, is an unsolved mystery. The objects have clearly suffered substantial orbital disturbances. The question is what caused them? Was it perhaps another passing star? For 2000 CR_{105} and Sedna that is feasible, at least in theory. They may have once been a part of the 'normal' scattered disk. If they were disturbed by a neighboring star at the furthest point in their orbits, it is possible that the *closest* point of those orbits were forced out to a greater distance. But Buffy's orbit cannot be explained in this way. It is almost circular, with its furthest point 'only' 9.2 billion kilometers from the Sun (61 AU). If a star ever passed at such a short distance, countless 'normal' Kuiper Belt objects would also have been seriously affected.

Brett Gladman of the University of British Columbia in Vancouver therefore thinks that the extended scattered disk was created not by the gravitational disturbances of passing stars, but by one or more planets in the outer regions of the solar system. If, in the early days of the Kuiper Belt, hundreds of Pluto-like objects were created (Alan Stern's thousand Plutos), it is not impossible that there were also a few really large objects among them, perhaps as massive as Mars or the Earth: 'ice giants' many thousands of kilometers across.

Gladman's computer simulations, published in *The Astrophysical Journal* in June 2006, show that you can explain the existence of the extended scattered disk with one or two of these troublemakers. Because of the gravitational effects of such a massive ice planet, Kuiper Belt objects whose motion has first been affected by Neptune end up in very elongated, highly inclined orbits, with their closest point at more than 6.5 billion kilometers from the Sun. Even extreme orbits like that of Sedna can be created in this way.

So where have all these ice planets gone to? The calculations by Gladman and his colleague Collin Chan also offer an answer to this: like the large majority of *trans*-Neptunian objects, they were expelled from the solar system by Neptune's gravitational effects between 100 and 200 million years after they were formed. It is possible that a few of these stray planets are still drifting around somewhere in the extended scattered disk, but most will certainly be at much greater distances away. Perhaps as far away as the Oort Cloud, like Stern's 1000 Plutos.

There, many billions of kilometers from the Sun and the Earth, in the cold, empty darkness of interstellar space, they now lead a retired existence, out of sight of earthly telescopes for ever. Technically speaking they may fulfill the IAU's definition of a planet, but no one will ever be able to observe or study them. No other objects in the solar system deserve the name 'Planet X' as much as these ice worlds, which lie beyond our reach forever.

The much smaller objects outside the orbit of Neptune – the ice dwarfs that have provided so many new insights on the early evolution of the solar system – will, however, soon reveal their final secrets. Cameras and detectors on board an unmanned space probe, racing toward the outer edge of the solar system at more than 20 kilometers a second, will study and photograph them from close up. At that speed, you can reach Pluto within 10 years and a few years later, you will cross the Kuiper Belt.

Pluto is being charted, ice dwarfs studied and measured. Alan Stern and his colleagues will soon be looking at new horizons. All we need is a little patience.

Chapter 30
New Horizons

New Horizons flies past Jupiter at a speed of 20 kilometers per second. The space probe photographs volcanic eruptions on Jupiter's moon Io, maps cloud swirls in the giant planet's atmosphere, and discovers interesting density clumps in its tenuous dust ring. It is February 28, 2007 and all the cameras and measuring instruments on board the probe are fully operational. But the passage does not last long. Its velocity considerably increased and its trajectory slightly changed by Jupiter's enormous gravitational pull, New Horizons continues its journey toward the distant outer regions of the solar system.

The Jupiter flyby was perfect and the results exceeded all expectations. The first close-ups were made of the Little Red Spot, a cloud formation that did not even exist when the space probe Cassini passed Jupiter at the end of 2000. New Horizons also witnessed an enormous eruption of the sulfur volcano Tvashtar on Io, which had remained dormant throughout many years of observation by the probe Galileo. New Horizons can rightly call itself the seventh Jupiter probe. Its flying visit to the giant planet was, however, only a short intermission on the way to its main destination, which it will not reach until July 14, 2015.

Artist impression of NASA's New Horizons spacecraft at Pluto (Johns Hopkins University Applied Physics Laboratory/Southwest Research Institute [JHUAPL/SwRI])

G. Schilling, *The Hunt for Planet X*, DOI 10.1007/978-0-387-77805-1_30,
© Springer Science+Business Media, LLC 2009

New Horizons is on the way to Pluto. The dwarf planet is to receive its first visit from a space probe. Eighty-five years after it was discovered, Pluto will reveal its secrets. What does its surface look like? Does it have mountains, cliffs, impact craters? What is the chemical composition of its frozen ground? Or of its thin atmosphere? And how is the atmosphere affected by electrically charged particles in the solar wind? The Jupiter flyby was only a dress rehearsal for the première in the summer of 2015, when Pluto will finally show its face.

The idea for a visit to Pluto dates back to May 1989. Twelve planetary scientists were eating pizza and drinking red wine at a small Italian restaurant in Baltimore after a conference on what was then the furthest planet, and they cooked up the idea for an exploratory flight. They included Marc Buie, Bob Millis, and Larry Wasserman from Lowell Observatory, asteroid expert Richard Binzel, and planetary satellite expert Bill McKinnon. And of course Alan Stern of the Southwest Research Institute in Boulder and his colleague Fran Bagenal of the University of Colorado. They called themselves the 'Pluto Underground' and were determined to persuade NASA of the need for a mission to Pluto.

That proved not to be too difficult at first. On August 25, 1989, NASA's space probe Voyager 2 flew past Neptune and made spectacular photographs of the large ice moon Triton. Triton's surface proved to be much more interesting than had always been believed, with nitrogen geysers and frozen ice lakes. And even though the existence of the Kuiper Belt had not yet been confirmed, it was generally assumed that Pluto and Triton would be very similar. A mission to the small, outer planet would therefore undoubtedly produce a wealth of interesting scientific results.

Geoff Briggs, director of NASA's Solar System Exploration Division, was therefore receptive to the Pluto Underground's ideas. At the end of 1989 he put Robert Farquhar of NASA's Goddard Space Flight Center in charge of a design study for 'Pluto 350' – a small space probe that would weigh no more than 350 kilograms. It would be relatively cheap to launch such a light probe. With flybys of Venus, the Earth, and Jupiter, Pluto 350 would gather sufficient velocity to reach Pluto within 15 years. Underground members Stern and Bagenal were appointed to the program to prepare the scientific aspects of the mission.

Pluto Fast Flyby

It all looked very promising. Early in 1991 Farquhar, Stern, and Bagenal submitted a detailed proposal to NASA's Solar System Exploration Subcommittee. Not long after, a special Outer Planets Science Working Group was set up, comprising not only many members of the Pluto Underground but also planetary scientists like Dale Cruikshank and David Tholen. They were a dedicated team, determined to make sure the plans became reality. No wonder then that they were not keen on NASA's proposal to launch a much larger Pluto probe weighing several tons. This mission, Mariner Mark II, was comparable to the Saturn probe Cassini and would cost some 2 billion dollars. The group found it much too ambitious, and therefore too risky.

But that did not apply to an unexpected competitor for Pluto 350: the Pluto Fast Flyby Project, designed by technicians Robert Staehle and Stacy Weinstein of the Jet Propulsion Laboratory in Pasadena. Staehle and Weinstein had been inspired by a postage stamp. In 1991, the US mail service had issued a series of nine stamps on the planets, with each stamp showing the most successful US probe to visit the planet. The stamp for Pluto, however, showed only a vague sphere and the text '*Not yet explored.*' High time to change that, thought Staehle and Weinstein. Their project was very cheap: with a mini-probe of 140 kilograms, which could reach Pluto in less than 8 years.

Faced with two proposals for a mission to Pluto, the Outer Planet Science Working Group (under Alan Stern's chairmanship) had to make a reasoned choice. The expected scientific impact of course played a role: the smaller and lighter a probe, the fewer cameras and measuring devices it can take along. But the fact that Pluto Fast Flyby was cheaper was a significant advantage in the early 1990s, when the NASA budget was increasingly under pressure. And so it happened, in 1992, that the plan for Pluto 350, to which working group chairman Alan Stern had devoted so much time and energy, was definitely shelved.

The choice for Pluto Fast Flyby by no means meant that the project was immediately given the green light. Many planetary scientists thought it irresponsible to spend so much money on such a revolutionary small space probe. It would not be possible to build on the technological experience acquired with previous projects and the scientific benefits would be marginal. Luckily Robert Staehle succeeded in interesting NASA's brand-new boss Daniel Goldin in the project. Pluto Fast Flyby fitted perfectly in Goldin's new 'faster, better, cheaper' philosophy.

Thanks to Goldin's involvement, the development of Pluto Fast Flyby moved into the fast lane, and the project received a great deal of publicity. During 1993 and 1994, miniature versions of thruster rockets, communications equipment, and scientific instruments were designed. If backups had to be made of all of these components (good standard practice in space exploration), the probe would be much too heavy. As an alternative, the design team decided to build two identical, lightweight probes – a much cheaper solution. That also gave the additional advantage that the two probes could pass Pluto and Charon on different sides, providing a more complete picture of the two objects.

Early in 1994, Alan Stern succeeded in interesting Russian scientists in the project. The Russian space research institute IKI agreed to build two small capsules which would make a descent into Pluto's thin atmosphere. And the two probes could be launched using cheap Russian Proton rockets. In 1995 Staehle and his colleagues came up with a new revolutionary design that weighed only 75 kilograms and used less energy than a light bulb. Because there was little resemblance to the original Pluto Fast Flyby, the project's name was changed to Pluto Express.

A Fresh Start

Despite all the technical progress, funding remained a problem. The estimated costs of the project continued to rise. After several design rounds, the backup probe had to be abandoned, but NASA still found the whole project too expensive. Even the fact that the Pluto probe (renamed the Pluto Kuiper Express) could also fly past one or two ice dwarfs in the Kuiper Belt at no great extra cost made little difference. It was by now the end of the 1990s and time was starting to run out. After the beginning of 2006 it would no longer be possible to use Jupiter to increase the velocity of a probe to Pluto, so a decision had to be made without delay.

That decision came in the fall of 2000. NASA announced that the flight to Pluto – by then budgeted at more than a billion dollars – would be scrapped, despite the fact that more than 200 million dollars had already been spent on design studies and hardware development. But the space agency had clearly not bargained for Pluto's enormous popularity among the general public. Within a day Ted Nichols, a high school student from Pennsylvania, had set up a website to campaign against NASA's decision and 2 weeks later more than 10,000 protest letters had been collected online.

NASA was impressed. Perhaps they had acted too soon. But maybe the project didn't need to be run by the agency's own Jet Propulsion Laboratory; instead they could put it out to tender among space research companies and institutes. Competition might produce what NASA itself had been unable to do – an affordable plan. At the end of 2000 NASA's director of Space Science Ed Weiler announced that the mission to Pluto might go ahead after all, on condition that NASA received fully detailed proposals within 3 months for a project that didn't cost a cent more than 500 million dollars.

Five teams succeeded in putting together serious plans within such a short period. One of them, New Horizons, was the brainchild of Alan Stern, who had sought the cooperation of the Applied Physics Laboratory at Johns Hopkins University in Laurel, Maryland. Space scientists and technicians at Johns Hopkins had previously been successful with the space probe NEAR-Shoemaker, which had orbited the asteroid Eros and, on February 12, 2001, had even made a semi-soft landing on its surface.

On June 6, NASA announced that two of the five proposals would receive funding for further study: New Horizons and Pluto and Outer Solar System Explorer (POSSE), a similar project led by Larry Esposito of the University of

Colorado in Boulder. To Stern's great relief, the decision meant that one of New Horizons' main competitors was out of the race: a plan for a Pluto probe with a revolutionary ion-drive engine. The plan had been submitted by a team led by Larry Soderblom of the United States Geological Survey in Flagstaff and was widely considered to be one of the leading contenders. At the end of September Stern and Esposito presented their detailed plans and, on November 29, NASA announced that it had chosen New Horizons.

Stern and his colleagues were in their seventh heaven, but the euphoria did not last long. To everyone's dismay, in its draft budget for 2003, presented to US Congress in February 2002, NASA had not reserved a cent for New Horizons. The Office of Management and Budget thought it irresponsible to give the go ahead for a cheap Pluto probe, largely because earlier proposals had led to enormous cost overruns. Once again, the plans were shelved, while the intended launch date (January 2006 at the latest) came closer and closer. Fortunately, during 2002, an intensive lobbying campaign, supported by a wave of articles in the popular media and a resounding recommendation from the National Research Council, succeeded in turning the tide. By the start of 2003, almost 14 years after the Pluto Underground had been set up, all potential obstacles had finally been cleared out of the way.

Souvenirs from Earth

New Horizons is a relatively small space probe weighing 481 kilograms. Its most striking feature is the 2.1-meter dish antenna for radio contact with the Earth. Beneath the dish is a triangular body containing seven scientific instruments. Together, the instruments weigh less than 30 kilograms and use only 30 Watts of energy. There are no solar panels – at distances of billions of kilometers from the Sun solar energy is not an option. The probe's energy is derived from a thermoelectric generator which utilizes the radioactive decay of plutonium.

But New Horizons has more on board than cameras, spectrometers, plasma instruments, and electronics. It also carries an American flag and a new Florida quarter, depicting the state as the Gateway to Discovery. There is a small fragment of carbon-fiber material from SpaceShipOne, the revolutionary manned spacecraft designed by space travel pioneer Burt Rutan which won the Ansari X Prize on October 4, 2004, paving the way for future commercial space tourism. And a CD-ROM with 430,000 names collected from throughout the world through the *Send your name to Pluto* Internet campaign.

New Horizons also pays tribute to the two people who, in a certain sense, the world has to thank for Pluto. An instrument for measuring interplanetary dust particles, developed by students at the University of Colorado, is officially known as the Venetia Burney Student Dust Counter, after the 11-year-old girl

who proposed the name Pluto for the dwarf planet in 1930. And there is a
capsule on board containing some of the ashes of Clyde Tombaugh, the man
who discovered Pluto. Tombaugh would have celebrated his 100th birthday
shortly after New Horizons was launched.

Alan Stern with Venetia Phair-Burney (Courtesy Alan Stern)

On Thursday, January 19, 2006, after a few nerve-wracking days during
which the launch was repeatedly delayed, New Horizons finally set off on its
several billion kilometer journey to Pluto and the Kuiper Belt. Fifty-eight
weeks later, the fastest spacecraft ever made completed its Jupiter flyby. But
then its long journey through empty space really started: another 100 months,
most of which the probe would spend in electronic hibernation. On Tuesday,
July 14, 2015, New Horizons will fly past Pluto at a distance of 10,000 kilometers and a relative velocity of 14 kilometers per second. During the flyby, it
will also study the large moon Charon and the small ice satellites Nix and
Hydra.

By coincidence, earlier that year another dwarf planet will receive a visitor
from Earth: the American space probe Dawn will study the asteroid Ceres
from close-up by. And even earlier, in November 2014, the European comet
probe Rosetta will set a small lander on the surface of comet Churyumov-
Gerasimenko, which almost certainly originated in the Kuiper Belt. The study
of small celestial bodies in the solar system – from irregular clumps of ice to
full-fledged dwarf planets – will clearly be in the glare of the spotlights in the
next decade.

The launch of New Horizons (NASA)

A 100 years after Percival Lowell presented his theoretical ideas on Planet X, planetary scientists will be closely examining the first detailed photographs of the surface of Pluto. By then tens of thousands of Kuiper Belt objects will undoubtedly have been discovered and perhaps even a few extremely remote dwarf planets larger than Eris. The hunt for Planet X, the search for unknown celestial bodies in the outer regions of the solar system, has not led to the

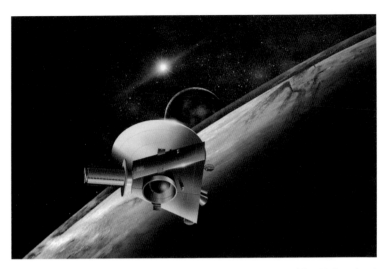

New Horizons at Pluto, with Charon in the background (Johns Hopkins University Applied Physics Laboratory/Southwest Research Institute)

discovery of just one *trans*-Neptunian object, but of a complete population of large and smaller ice dwarfs which provide a surprising new picture of the period when the planets were formed.

Planet X may not exist, but the hunt will never come to an end. There will always be new horizons.

Table 283

Table 1 Details of a number of large ice dwarfs

Name and/or nickname	Official number	Provisional designation (giving year of discovery)	Semi-major axis (billion km)	Perihelion distance (billion km)	Aphelion distance (billion km)	Eccentricity	Orbital period (year)	Orbital inclination	Estimated diameter (km)	Number of satellites	Type*
Pluto	134340	(discovered in 1930)	5.906	4.437	7.376	0.249	248.1	17.1°	2300	3 (Charon, Nix, Hydra)	P
Chiron	2060	1977 UB	2.045	1.263	2.826	0.382	50.54	6.9°	140	–	c
Pholus	5145	1992 AD	3.056	1.306	4.807	0.573	92.35	24.7°	180	–	c
'Smiley'	15760	1992 QB_1	6.543	6.115	6.970	0.065	289.2	2.2°	160	–	K
–	15874	1996 TL_{66}	12.402	5.240	19.563	0.577	754.8	24.0°	600	–	s
Varuna	20000	2000 WR_{106}	6.451	6.121	6.782	0.051	283.2	17.2°	900	–	K
Ixion	28978	2001 KX_{76}	5.936	4.501	7.371	0.242	250.0	19.6°	800	–	P
–	55565	2002 AW_{197}	7.074	6.143	8.004	0.132	325.2	24.4°	700	–	K
Quasar ('mini-Xena')	50000	2002 LM_{60}	6.493	6.270	6.716	0.034	286.0	8.0°	1250	1	K
Sedna ('Flying Dutchman')	90377	2003 VB_{12}	78.629	11.393	145.867	0.855	12,050.3	11.9°	1500	–	s
Orcus	90482	2004 DW	5.897	4.567	7.227	0.226	247.5	20.6°	950	1	K
'Buffy'	-	2004 XR_{190}	8.535	7.826	9.243	0.083	430.9	46.7°	750	–	K
'Santa'	136108	2003 EL_{61} (discovered in 2004)	6.484	5.260	7.708	0.189	285.4	28.2°	1500	2	K
Eris ('Xena')	136199	2003 UB_{313} (discovered in 2005)	10.123	5.65	14.60	0.442	557	44.2°	2400	1(Dysnomia)	s
'Easter Bunny'	136472	2005 FY_9	6.850	5.761	7.940	0.159	309.9	29.0°	1800	–	K

* c = centaur. κ = classical Kuiper Belt object. p = plutino. s = scattered disk object
Note: In 2008, the ice dwarfs 2003 EL_{61} ('Santa') and 2005 FY_9 ('Easter Bunny') were classified as dwarf planets, and given the names Haumea and Makemake, respectively. The two satellites of Haumea were given the names Hi'iaka and Namaka.

Chronology

• 1612	– Galileo Galilei observes Neptune without recognizing it as a planet
• 1682	– Edmund Halley predicts the return of the comet bearing his name in 1758
• 1690	– John Flamsteed observes Uranus and catalogs it as star 34 Tauri
• 5-8-1774	– Planetary conjunction which inspires Eise Eisinga to build his planetarium
• 3-13-1781	– William Herschel discovers Uranus
• 9-20-1800	– The *Himmelpolizei* meets in Lilienthal to determine a strategy for the hunt for a planet between the orbits of Mars and Jupiter
• 1-1-1801	– Giuseppe Piazzi discovers the first asteroid, Ceres
• 3-28-1802	– Heinrich Olbers discovers the second asteroid, Pallas
• 5-6-1802	– William Herschel suggests the name *asteroids* for the small celestial bodies between the orbits of Mars and Jupiter
• 9-1-1804	– Karl Harding discovers the third asteroid, Juno
• 3-29-1807	– Heinrich Olbers discovers the fourth asteroid, Vesta
• 3-11-1811	– Birth of Urbain Le Verrier
• October 1845	– John Adams tries in vain to present his calculations of the position of an unknown planet beyond the orbit of Uranus to Astronomer Royal George Airy
• 7-30-1846	– While searching for a planet beyond the orbit of Uranus, James Challis observes the planet Neptune without recognizing it as such
• 9-18-1846	– Urbain Le Verrier writes to Johann Galle predicting the position of an unknown planet beyond the orbit of Uranus

- **9-23-1846** — Johann Galle discovers Neptune
- **3-13-1855** — Birth of Percival Lowell
- **3-26-1859** — Edmond Lescarbault believes he has observed the transition of a intramercurial planet (Vulcan)
- **1-2-1860** — Urbain Le Verrier calls on his colleagues to hunt for Vulcan
- **7-18-1860** — Astronomers search in vain for Vulcan during a total eclipse of the Sun
- **4-4-1875** — German astronomers claim to have observed the transition of Vulcan
- **9-23-1877** — Death of Urbain Le Verrier
- **7-29-1878** — American astronomers believe they have observed Vulcan during a total solar eclipse
- **6-1-1894** — Lowell Observatory, where Pluto was discovered in 1930, is officially opened
- **12-7-1905** — Birth of Gerard Kuiper
- **2-4-1906** — Birth of Clyde Tombaugh
- **3-19-1915** — Pluto is photographed during a search for a planet beyond the orbit of Neptune, but is not recognized as such
- **11-12-1916** — Death of Percival Lowell
- **1-23-1930** — Clyde Tombaugh makes the photograph on which he discovers Pluto
- **2-18-1930** — Clyde Tombaugh discovers the dwarf planet Pluto
- **3-13-1930** — The discovery of Pluto is announced to the world
- **3-14-1930** — Venetia Burney suggests the name 'Pluto' for the newly discovered trans-Neptunian planet
- **5-1-1930** — The name 'Pluto' is officially announced
- **5-3-1931** — Première of the first Disney cartoon in which Mickey Mouse's dog is called Pluto
- **1936** — Raymond Lyttleton publishes his theory that Pluto was once a Neptunian moon
- **1938** — Birth of Jim Christy
- **1943** — Kenneth Edgeworth publishes his theory on a comet belt beyond the orbit of Neptune
- **1948** — Gerard Kuiper discovers Uranus' moon Miranda
- **1950** — Jan Oort publishes his theory on a comet cloud at a great distance from the Sun
- **1951** — Gerard Kuiper publishes his theory on a comet belt beyond the orbit of Neptune
- **1965** — Otto Franz photographs Pluto's moon Charon without recognizing it as such
- **6-5-1965** — Birth of Mike Brown

• 3-2-1972	– Launch of Pioneer 10
• 4-5-1973	– Launch of Pioneer 11
• 12-4-1973	– Pioneer 10 flies past Jupiter
• 12-23-1973	– Death of Gerard Kuiper
• 12-3-1974	– Pioneer 11 flies past Jupiter
• March 1976	– Dale Cruikshank, David Morrison and Carl Pilcher conduct the first spectroscopic observations of Pluto's surface
• 11-19-1976	– Dale Cruikshank, David Morrison and Carl Pilcher publish their spectroscopic observations of Pluto's surface in *Science*
• 8-20-1977	– Launch of Voyager 2
• 9-5-1977	– Launch of Voyager 1
• 10-18-1977	– Charles Kowal makes the photograph on which the first centaur, Chiron is discovered
• 11-4-1977	– The discovery of Chiron is announced
• 6-22-1978	– Jim Christy discovers Pluto's moon Charon
• 7-7-1978	– The discovery of Charon is announced
• 3-5-1979	– Voyager 1 flies past Jupiter
• 7-9-1979	– Voyager 2 flies past Jupiter
• 9-1-1979	– Pioneer 11 flies past Saturn
• 1980	– Alistair Walker observes a stellar occultation by Pluto's moon Charon
• 1980	– Julio Fernández publishes his theory that short-period comets originate in the Kuiper Belt
• 11-12-1980	– Voyager 1 flies past Saturn
• 8-27-1981	– Voyager 2 flies past Saturn
• 10-16-1982	– Dave Jewitt rediscovers Halley's Comet
• 1984	– Julio Fernández and Wing-Huen Ip publish their theory of planetary migration
• 4-19-1984	– The Nemesis theory of Richard Muller, Piet Hut and Marc Davis is published in *Nature*
• 1985	– First observation of mutual occultations and eclipses of Pluto and Charon
• 1-24-1986	– Voyager 2 flies past Uranus
• 1987	– Martin Duncan, Thomas Quinn and Scott Tremaine publish their computer simulations of the origins of the Oort Cloud
• 1988	– Martin Duncan, Thomas Quinn and Scott Tremaine publish their computer simulations of the evolution of short-period comets
• 6-9-1988	– First successful observation of a stellar occultation by Pluto
• May 1989	– First meeting of the 'Pluto Underground', where the idea of an unmanned mission to Pluto is born

- **8-25-1989**   – Voyager 2 flies past Neptune
- **April 1991**   – Alan Stern publishes his theory on the existence of 'a thousand Plutos'
- **1-9-1992**   – David Rabinowitz discovers the second centaur, Pholus
- **8-30-1992**   – Dave Jewitt and Jane Luu discover the first ice dwarf, 1992 QB$_1$
- **9-14-1992**   – The discovery of 1992 QB$_1$ is announced
- **Dec 1992**   – Pioneer 10 changes course slightly, possibly after an encounter with a small ice dwarf
- **March 1993**   – Dave Jewitt and Jane Luu discover the second ice dwarf, 1993 FW
- **September 1993**   – Dave Jewitt and Jane Luu discover the third and fourth ice dwarfs, 1993 RO and 1993 RP
- **9-16-1993 and 9-17-1993** – Iwan Williams and his colleagues discover the fifth and sixth ice dwarfs, 1993 SB and 1993 SC
- **10-28-1993**   – Renu Malhotra publishes her theory on the origins of Pluto's orbit in *Nature*
- **3-8-1996**   – Publication of the first 'world maps' of Pluto, on the basis of Hubble photos
- **10-9-1996**   – Discovery of the first 'scattered disk' object, 1996 TL$_{66}$
- **1-17-1997**   – Death of Clyde Tombaugh
- **1-30-1997**   – The discovery of 1996 TL$_{66}$ is announced
- **6-5-1997**   – The discovery of 1996 TL$_{66}$ is published in *Nature*
- **9-25-1997**   – Sigeru Ida, Robin Canup and Glen Stewart publish their computer simulations of the origins of the Moon in *Nature*
- **2-3-1999**   – The International Astronomical Union announces that Pluto's planetary status will not be revised
- **11-28-2000**   – Discovery of the large ice dwarf Varuna
- **122-2-2000**   – Christian Veillet and his colleagues discover that ice dwarf 1998 WW$_{31}$ is a binary object
- **1-22-2001**   – The *New York Times* writes about the demotion of Pluto in the American Museum for Natural History in New York
- **2-12-2001**   – The space probeNEAR Shoemaker lands on the asteroid Eros
- **4-16-2001**   – The discovery of the binary nature of ice dwarf 1998 WW$_{31}$ is announced
- **5-22-2001**   – Discovery of the large ice dwarf Ixion
- **August 2001**   – Discovery of the first Neptunian trojan

- **11-29-2001**   – NASA selects the space probe New Horizons for a mission to Pluto
- **1-10-2002**   – Discovery of the large ice dwarf 2002 AW_{197}
- **6-5-2002**   – Discovery of the large ice dwarf Quasar
- **7-20-2002**   – Second successful observation of a stellar occultation by Pluto
- **8-21-2002**   – Third successful observation of a stellar occultation by Pluto
- **10-7-2002**   – The discovery of Quasar is announced
- **3-7-2003**   – The photos are made on which Pablo Santos-Sanz discovers 2003 EL_{61}
- **10-21-2003**   – The photos are made on which Eris is discovered
- **11-14-2003**   – Discovery of the large ice dwarf Sedna
- **2-17-2004**   – Discovery of the large ice dwarf Orcus
- **2-19-2004**   – The discovery of Orcus is announced
- **3-15-2004**   – The discovery of Sedna is announced
- **5-6-2004**   – The photos are made on which Mike Brown discovers 2003 EL_{61}
- **6-19-2004**   – Discovery of the Earth-grazer Apophis
- **12-11-2004**   – The photos are made on which the extraordinary ice dwarf 2004 XR_{190} ('Buffy') is discovered
- **12-13-2004**   – The discovery of 2004 XR_{190} ('Buffy') is announced
- **12-28-2004**   – Mike Brown discovers the large ice dwarf 2003 EL_{61}
- **1-5-2005**   – Mike Brown discovers the dwarf planet Eris
- **1-28-2005**   – Robin Canup publishes her computer simulations of the origin of Pluto's moon Charon in *Science*
- **1-28-2005**   – Mike Brown and his colleagues discover a small satellite around the large ice dwarf 2003 EL_{61}
- **3-31-2005**   – Mike Brown discovers the large ice dwarf 2005 FY_9
- **5-15-2005 and 5-18-2005**   – The photographs on which the small Plutonian satellites Nix and Hydra are discovered are made by the Hubble Space Telescope
- **5-26-2005**   – Hal Levison, Alessandro Morbidelli, Rodney Gomes and Kleomenis Tsiganis publish their Nice model of the early evolution of the solar system in *Nature*
- **6-15-2005**   – Max Mutchler discovers the small Plutonian satellites Nix and Hydra
- **6-30-2005**   – Mike Brown and his colleagues discover a second satellite around the large ice dwarf 2003 EL_{61}

• **7-4-2005**	– The space probe Deep Impact passes close to comet Tempel 1 and shoots a projectile into the cometary nucleus
• **7-20-2005**	– Mike Brown and his colleagues publish the abstracts of their DPS presentations on the discovery of the large ice dwarf 2003 EL_{61} on the internet
• **7-25-2005**	– Pablo Santos-Sanz discovers the large ice dwarf 2003 EL_{61}, independently from Mike Brown
• **7-29-2005**	– The discovery of 2003 UB_{313} (Eris), 2003 El_{61} and 2005 FY_9 is announced
• **8-17-2005**	– Andrew Steffl discovers the small Plutonian satellites Nix and Hydra independently of Max Mutchler
• **9-10-2005**	– Mike Brown and his colleagues discover the small satellite Dysnomia around the dwarf planet Eris
• **9-30-2005**	– The discovery of the small satellite of Eris is announced
• **10-31-2005**	– The discovery of the small Plutonian satellites Nix and Hydra is announced
• **1-19-2006**	– Launch of the Pluto probe New Horizons
• **2-23-2006**	– The discovery of the small Plutonian satellites Nix and Hydra is published in *Nature*
• **2-28-2007**	– The Pluto probe New Horizons flies past Jupiter
• **3-13-2007**	– The State of New Mexico announces that Pluto is a planet
• **6-12-2006**	– Fourth successful observation of a stellar occultation by Pluto
• **6-20-2006**	– The names of the small Plutonian satellites Nix and Hydra are officially announced
• **6-30-2006**	– The seven-man planet definition committee of the IAU meets in Paris
• **8-16-2006**	– The IAU presents the planet definition committee's 'twelve-planet proposal'
• **8-24-2006**	– The IAU decides that Pluto is no longer a genuine planet, but a dwarf planet
• **9-6-2006**	– Pluto is allocated 'asteroid number' 136199
• **9-13-2006**	– The names of the dwarf planet Eris and its satellite Dysnomia are officially announced
• **3-15-2007**	– Mike Brown and his colleagues publish their discovery of collisional fragments of the large ice dwarf 2003 EL_{61} in *Nature*

Glossary of terms

Adaptive optics A technology used in telescopes to compensate for the effects of atmospheric turbulence

Albedo Measure of the extent to which a body reflects light

Aphelion Point in the elliptical orbit of a solar system object at which it is farthest from the Sun

Arc minute Sixtieth part of a degree

Arc second Sixtieth part of an arc minute; 3,600th part of a degree

Asteroid Small, mostly rocky celestial body orbiting the Sun, usually within the orbit of Jupiter

Astronomical unit Average distance between the Earth and the Sun: 149.6 million kilometers (92.96 million miles)

Binary planet Planet with a relatively large companion; the Earth and the Moon are sometimes described as a binary planet

Blink comparator An instrument with which two photographic plates can be compared, used in the past to search for faint, moving points of light

CCD Charge coupled device: light-sensitive chip comprising a large number of individual pixels

Centaur Small, icy body in an orbit around the Sun which lies largely between the orbits of Saturn and Neptune

Classical Kuiper Belt object Ice dwarf in a more or less circular orbit which lies entirely beyond the orbit of Neptune and having no orbital resonance with the giant planet

Comet Small, icy body in an elongated orbit around the Sun, which can exhibit a striking tail of gas and/or dust as a consequence of solar radiation

Conjunction Apparent proximity of two celestial bodies in the sky

Cubewano Little used name for classical Kuiper Belt objects (derived from 1992 QB$_1$)

Degree Ninetieth part of a right angle; 360th part of a circle

Density Average amount of material per unit volume; quotient of the mass and volume of a celestial body

Differentiation Process in which heavier materials sink towards the core of a celestial body, while the lighter materials float towards the surface

Dwarf planet Celestial body in an orbit around the Sun which is in a state of hydrostatic equilibrium (and therefore large enough for their own gravity to have formed them into a spherical shape), but is not dynamically dominant in its own orbital region

Earth-grazer Asteroid that may pass relatively close to the Earth

Eclipse Event that occurs when a celestial body casts its shadow on another object

Exoplanet Planet orbiting another star than the Sun

Ice dwarf Small object orbiting the Sun in the outer regions of the solar system which consist largely of ice

Interplanetary space Space between the planets

Interstellar space Space between the stars

Intramercurial planet Hypothetical planet within the orbit of Mercury

Kuiper Belt Broad belt of ice dwarfs in the outer regions of the solar system, named for the Dutch-American planetary scientist Gerard Kuiper

Kuiper Belt object Ice dwarf in the Kuiper Belt (see also: trans-Neptunian object)

Lightyear Distance traveled by a ray of light in one year: 9.46 trillion kilometers (5.88 trillion miles)

Long-period comet Comet with an orbital period of more than 200 years

Migration Process in which the orbit of a celestial body becomes larger or smaller, for example through the gravitational influence of other bodies or a disk of gas and dust

Minor planet General term for asteroids, ice dwarfs and dwarf planets

Moon Small object in orbit around a larger body that is orbiting the Sun

Occultation Event that occurs when a celestial body is hidden from view by another object passing in front of it

Oort cloud An extended, more or less spherical cloud of trillions of small comets at a great distance from the Sun, named for the Dutch astronomer Jan Oort

Opposition Event that occurs when a solar system body is exactly opposite the Sun in the sky (as seen from the Earth)

Orbital angular momentum Measure of the kinetic energy of a celestial body in orbit around another object

Orbital eccentricity Measure of the elliptical shape of the orbit of a celestial body.

Orbital elements Six properties (including orbital inclination and orbital eccentricity) which fully characterize the shape, size and orientation of the orbit of a celestial body

Orbital inclination Angle between the orbital plane of a celestial body and that of the Earth

Orbital resonance Situation in which the orbital periods of two celestial bodies exhibit a simple ratio (like 2:1 or 3:2) due to a periodic gravitational influence on each other

Perihelion Point in the elliptical orbit of a solar system object where it is closest to the Sun

Planet Celestial body in orbit around the Sun, which is in hydrostatic equilibrium (and therefore large enough for their own gravity to have formed them into a spherical shape) and dynamically dominant in its own orbital region

Planetesimal Small lump of rock and/or ice created by aggregation in a protoplanetary disk; many planetesimals later merged to form protoplanets

Plutino Kuiper Belt object which, like Pluto, has a 3:2 orbital resonance with Neptune

Plutoid Dwarf planet in the Kuiper Belt

Protoplanet Celestial body orbiting the Sun formed during the early stages of the solar system by small planetesimals merging as a consequence of their mutual gravitational influence

Protoplanetary disk Flat, rotating disk of gas and dust around a new-born star from which planets can be formed

Resonance See orbital resonance

Satellite See moon

Scattered disk Very extended region in the outer solar system containing a large number of ice dwarfs expelled from the Kuiper Belt by the gravitational influence of Neptune and other objects

Short-period comet Comet with an orbital period of less than 200 years

Solar nebula Protoplanetary disk around the sun from which the planets were formed

Stellar occultation Event that occurs when a celestial body, for example an ice dwarf, passes in front of a remote star

Solar system Collective name for the Sun and all the large and smaller objects orbiting it

Spectrum Graph showing the brightness of a celestial body against wavelength; the exact form of the spectrum allows the surface composition of a celestial body to be determined

Survey Astronomical observing program entailing a systematic search for a specific kind of object in a large part of the night sky

Transit Event that occurs when a small celestial body moves across the face of a larger body

Trans-Neptunian object Small, icy celestial body in an orbit around the Sun that is mostly beyond the orbit of Neptune (see also: Kuiper Belt object)

Trojan Small, icy celestial body in a more or less stable orbit around the Sun, whose orbital period is identical to that of a large planet

Vulcanoid Hypothetical asteroid orbiting the Sun within the orbit of Mercury

Zodiac Belt of twelve constellations in the sky in which (seen from the Earth) the planets are all located

Bibliography

The Search for Planet X, Tony Simon, Scholastic Book Services, 1965

Planets X and Pluto, William Graves Hoyt, University of Arizona Press, 1980

Out of the Darkness: The Planet Pluto, Clyde Tombaugh, Signet, 1981

Clyde Tombaugh and the Search for Planet X, Margaret Wetterer and Laurie Caple, Carolrhoda Books, 1996

Dark Matter, Missing Planets and New Comets: Paradoxes Resolved, Origins Illuminated (2nd edition), Tom Van Flandern, North Atlantic Books, 1999

Beyond Pluto, John Davies, Cambridge University Press, 2001

Minor Bodies in the Outer Solar System: Proceedings of the ESO Workshop Held at Garching, Germany, 2–5 November 1998, Alan Fitzsimmons, Dave Jewitt and Richard West (eds.), Springer, 2001

The First Decadal Review of the Edgeworth-Kuiper Belt, John Davies and Luis Barrera (eds.), Springer, 2004

Planets Beyond: Discovering the Outer Solar System, Mark Littmann, Dover, 2004

Pluto and Charon: Ice Worlds on the Ragged Edge of the Solar System (2nd edition), Alan Stern and Jacqueline Mitton, Wiley-VCH, 2005

Is Pluto a Planet? A Historical Journey through the Solar System, David Weintraub, Princeton University Press, 2006

Pluto and Beyond: A Story of Discovery, Adversity, and Ongoing Exploration, Anne Minard, Northland Publishing, 2007

Clyde Tombaugh: Discoverer of Planet Pluto, David Levy, Sky Publishing, 2007

Pluto: From Planet to Ice Dwarf, Elaine Landau, Children's Press, 2007

The Pluto Files: The Rise and Fall of America's Favorite Planet, Neil deGrasse Tyson, W.W. Norton, 2009

Index

Made in the USA
Lexington, KY
09 May 2011